Genetics in Clinical Practice

New Directions for Nursing and Health Care

Dale Halsey Lea, RN, MPH
Foundation for Blood Research
Southern Maine Regional Genetics Service
Scarborough, Maine

Jean F. Jenkins, RN, MSN, PhD(c)
National Human Genome Research Institute
Public Health Service
Bethesda, Maryland

Clair A. Francomano, MD
National Human Genome Research Institute
Bethesda, Maryland

JONES AND BARTLETT PUBLISHERS
Sudbury, Massachusetts
BOSTON   TORONTO   LONDON   SINGAPORE

World Headquarters
Jones and Bartlett Publishers
40 Tall Pine Drive
Sudbury, MA 01776
800-832-0034
978-443-5000
info@jbpub.com
www.jbpub.com

Jones and Bartlett Publishers Canada
P.O. Box 19020
Toronto, ON M5S 1X1
CANADA

Jones and Bartlett Publishers International
Barb House, Barb Mews
London W6 7PA
UK

Copyright © 1998 by Jones and Bartlett Publishers, Inc.

Production Credits
Sponsoring Editor: Karen McClure
Production Editor: Lianne Ames
Manufacturing Buyer: Jane Bromback
Design: Publicom, Inc.
Editorial Production Service: Publicom, Inc.
Illustration: Publicom, Inc.
Typesetting: Publicom, Inc.
Cover Design: Dick Hannus
Printing and Binding: Malloy Lithographing
Cover Printing: Malloy Lithographing

Library of Congress Cataloging-in-Publication Data

Lea, Dale Halsey.
    Genetics in clinical practice : new directions for nursing and
  health care / Dale Halsey Lea, Jean F. Jenkins, Clair A. Francomano.
      p. cm.
    Includes bibliographical references and index.
    ISBN 0-7637-0542-X
    1. Medical genetics. 2. Nurses. 3. Allied health personnel.
I.   Jenkins, Jean F. II. Francomano, Clair A. III. Title.
    [DNLM: 1. Genetic Technique nurses' instruction. 2. Gene Therapy
  nurses' instruction. 3. Genetic Counseling nurses' instruction.
QZ 50 L433g 1998]
RB 155.L382 1998
616'. 042-dc21
DNLM/DLC
for Library of Congress   98-1421
                                           CIP

Cover art: The Genetics Quilt shown on the cover is used with permission
of The Sarum Quilters, Salisbury Health Care NHS Trust, and Peter Read, photographer.

ISBN 0-7637-0542-X

Printed in the United States of America
02 01 00 99 98   10 9 8 7 6 5 4 3 2 1

The primary aim of all education must be nurturance of the ethical ideal.

Nel Noddings

This book is dedicated to friends, families, and associates, and especially to

Jessie Parkes Halsey
Sophia Mumford
Carolyn Fish
Hazel Johnson Brown
Sue Hubbard
Ruth Malden Francomano

CONTENTS

CHAPTER 1

The Human Genome Project   1

CHAPTER 4

Integrating Genetics into Nursing Practice   71

CHAPTER 5

What to Expect from Genetic Counseling   109

CHAPTER 6

The Process of Genetic Testing for Nurses and Their Patients   145

CHAPTER 7

The Genetic Basis of Cancer   177

CHAPTER 8

Cancer-Risk Assessment and Counseling   193

CHAPTER 9

Ethics, Genetics, and Nursing Practice   221

A significant expansion of scientific knowledge in human genetics has occurred during the last decade. Due in large part to efforts of the Human Genome Project, many disease-related genes have been identified, leading, in turn, to the rapid development of genetic tests. The speed with which new findings are translated into clinical applications is astounding. As more genetic technologies and information become available to the public, nurses will be increasingly called upon to participate in discussing and interpreting genetic information and to assist patients, families, and the broader community in understanding, assimilating, and adjusting to new genetic information.

Many exciting initiatives are under way to integrate new genetic knowledge and skills into clinical practice and to promote coordination of educational activities in genetics among all health professionals. The overall goal of these efforts is to increase health professionals' understanding of human health and disease and to ensure high-quality care and services for patients, families, and communities who may have genetic health conditions that affect their present or future health. For this goal to be achieved, all health professionals must also have an understanding of the ethical, legal, and social impact of genetic advances.

Nurses have a long tradition of caring for patients and families who have genetic conditions and, in collaboration with other health professionals, have taken a leading role in assuring coordination and continuity of care. Incorporating new genetic knowledge into the scope and standards of practice will help nurses to continue to contribute to health promotion, advocacy, and patient support.

Genetics in Clinical Practice: New Directions for Nursing and Health Care was written in response to the expressed need for collaborative educational efforts in genetics for all health professionals. This book is intended to provide nurses with a basic understanding of genetic advances and their applications in order to support and enhance nursing practice in genetics. It builds on past collaborative genetics educational endeavors for nurses and will contribute to future efforts to increase nurses' knowledge and understanding of genetics and its role in the wellness-illness continuum and in health care more generally.

Elizabeth Thomson, RN, MS, Director, Clinical Genetics Research
    Ethical, Legal, and Social Implications Research Program
National Human Genome Research Institute
National Institutes of Health

Francis Collins, MD, PhD, Director, National Human Genome
    Research Institute
National Institutes of Health

No matter what skills or tasks are performed by the nurse, the salient dimension is the underlying beliefs and values that are brought to bear on how skills are used and tasks accomplished. Self-awareness, centeredness, and caring suggest an internal set of principles underlying this philosophy.

Mary B. Johnson, 1990

During the last 40 years, the evolution of scientific discoveries has caused an explosion of knowledge that has had tremendous influence on available health care options. Understanding the role that genetics plays in health and disease is opening up new opportunities for diagnosis, prevention, and treatment of many common diseases.

This book focuses on the expanding impact of genetic technologies on nursing practice and the challenge of integrating genetics into nursing practice. What role will you play in moving these discoveries from the research arena to the general community? This book will give you a foundation on which to build in responding to the rapid discovery of genes and to understanding their consequences. In providing patient care in the future, you will face many challenges, concerns, and controversial issues. This book offers a starting point from which you can begin planning for the educational, research, and clinical interventions needed to support patients and their families successfully. As the secrets of the human genome are unlocked, nurses will be required to become actively involved in the clinical application of this sensitive and personal information.

Nurses, regardless of their specialty area, role, or practice setting, will be faced with questions about genetics. Questions from individuals and families at work, neighbors, and friends, and even personal concerns will become more prevalent as more is learned about human genetics. As educators, nurses may be asked to explain how human gene function affects human development and variation.

Translating these discoveries into practical information for individuals, society, and the health care delivery system will challenge each of us. At no time has the need been more immediate for nurses and other health care professionals to become better informed about developing genetic technologies. This knowledge can be used to design health care resources and systems to meet the future needs of patients, families, and the broader community. Exploration of values, beliefs, and knowledge about how to use genetic information is therefore essential to be able to build the infrastructure effectively and to provide resources. Only then will society and the health care community be able to achieve equitable and effective application of this powerful technology.

Knowledge of each individual genetic blueprint encompasses the concepts of human genetic similarities and personal uniqueness. This knowledge has the potential to challenge long-held personal values, beliefs, and understanding of how the human body develops and functions. New scientific explanations will facilitate development of options for health care and may create personal and professional dilemmas related to the use of this sensitive information. Nurses will need to open their minds to the possibilities ahead and become aware of how their personal beliefs affect application of this knowledge in their professional role.

Self-awareness contributes to the development of quality nursing care that incorporates genetics in several ways:

- A nurse's personal understanding of views on the application of genetic technology may affect the selection of and manner in which information is told to a patient, which may, in turn, influence patient decision making and affect outcomes. Self-awareness is also essential if nurses are to be able to integrate current genetic technology with ethical decision-making processes. Understanding that patients' health decisions are based on past experiences, values, interests, motivations, and expectations is integral to this process.
- The rapidly changing health care environment may create uncertainty and role ambiguity that will influence the confidence and competence of nurses in providing quality patient care. The ability to develop ethical competence related to genetics requires time, skills, and self-awareness. Preparation of nurses through profes-

sional ethics education facilitates the ability of nurses to integrate genetics information into clinical decision making and to avoid struggling with ethical uncertainties born of new technologies.

• Balancing technological innovation with responsiveness and sensitivity to concerns of patients and families is an important nursing contribution. The views of nurses will influence how such information is used and interpreted to patients. Becoming aware of one's beliefs, reservations, and concerns about progress in understanding the genetic contribution to health and illness is a starting point for dialogue between the professional nurse and the community. Through enhanced communication and discussion about the implications of genetic technology, nurses can begin to understand the significance of these discoveries, both personally and professionally. Clarification of the questions, concerns, and needs of the community in understanding genetics will provide a foundation on which nurses can build options that are of value for the health care of the future.

This book is intended to introduce nurses to genetic aspects of health care and to support the integration of genetic concepts into nursing practice. The book provides concepts of basic human genetics in the context of major genetic discoveries and emerging ethical, legal, and potential social issues surrounding the application and use of genetic technologies. It addresses current and future applications of genetics to nursing practice.

Nurses in undergraduate programs may find this book useful, as it is aimed at preparing nurses to integrate genetics into care that is planned and provided to persons, families, or communities in all practice settings. It supports nursing competencies in assessment of genetic health, communication of genetic information, utilization of genetic technology, human valuing, coordination of care, and ethical reasoning and critical thinking. It may also be used as a resource for continuing-education courses in community, maternal, and child health, for medical and surgical nursing, and for other allied health professionals who may collaborate with nurses in providing care.

Each chapter is preceded by one or more learning objectives, a rationale for the objectives, and anticipated practice activities. The book begins with a self-assessment exercise that will help guide the nurse through exploration of personal beliefs, values, and practice in genetics. Case studies are used throughout to explore practice-based applications and integration of genetic information. A text or summary article is referenced for each genetic condition described, and family and community genetics resources are included to enhance nursing partnership with patients and families. Additional genetic and nursing references are listed at the end of each chapter. Questions for critical thinking are included for self-learning and discussion and to promote further integration of genetic concepts, principles, and applications. A glossary of terms, listings of current policy statements on genetic topics, and selected professional nursing and genetic resources are also provided at the end of the book.

Chapters 1–3 present an overview of the Human Genome Project, genetic trends and discoveries, and basic human genetics concepts. It is our hope that this information will create a global health care perspective and provide a framework around which readers may build awareness and a broad understanding of genetics in health care and daily living.

Chapters 4 and 5 discuss integration of genetic advances into nursing practice, including participation in the genetic counseling process. Chapter 5 describes the components of genetic counseling, related issues and nursing activities throughout the health care continuum that will help nurses prepare patients to participate in the counseling process. Chapter 6 focuses on genetic testing and provides case studies and in-depth discussions that outline nursing issues and roles. Chapters 7 and 8 summarize some of the new advances in our understanding of cancer genetics and emerging counseling issues and discuss nursing participation in this new field of predisposition genetic testing. Chapter 9 reviews some of the ethical and social issues of predisposition genetic assessment, such as the value and importance of informed choice, the confidentiality of genetic information, and the potential for discrimination; explores these and other ethical and social issues; and discusses nursing applications. Chapter 10 describes future issues in genetics and encourages nurses to look ahead to trends in genetics study and the design of health care.

Appendix A consists of three case studies that provide an opportunity for further discussion and critical thinking. Other appendixes include a current listing of selected social policy statements regarding genetic issues (e.g., privacy and adoption) and professional and patient resources for nurses.

Nurses have a unique and holistic view of patients, families, and communities that takes into account physical, social, psychological, and spiritual responses to health and illness. Nurses also care for patients throughout the life span and collaborate with a variety of health professionals to ensure continuity and coordination of care that is supportive of health promotion and healthy adaptation to illness. We hope that this book will advance the integration of genetics principles and concepts into nursing practice and contribute to the development of knowledgeable nurses who are prepared to provide quality genetic nursing care.

Jean Jenkins

Clair A. Francomano

ACKNOWLEDGEMENTS

We gratefully acknowledge the support and assistance of Francis Collins; Elizabeth Thomson; Kathleen Calzone; the Foundation for Blood Research; the International Society of Nurses in Genetics; Thomas, Allison, Halsey, and Annie Lea; Debbie and Chandler Hawkes; Jamie and Jeremy Jenkins; Brenda Hildebrand; Betty Wolf-Ledbetter; M. F. D'Loring; and Elizabeth Garabedian.

CONTRIBUTORS

Evan DiRenzo, PhD
Senior Staff Bioethics Fellow
National Institutes of Health, Clinical
  Center
Bethesda, Maryland

Eileen Diamond RN, MS
Cancer Genetics Nurse
NCI - Medicine Branch at Navy
Bethesda Naval Hospital
Bethesda, Maryland

Christine Grady, PhD, RN, FAAN
Department of Clinical Bioethics
National Institutes of Health

David H. Ledbetter, PhD, FACMG
Director, Center for Medical Genetics
University of Chicago
Chicago, Illinois

June Peters, MS, CGC
Genetic Counselor
National Human Genome Research
  Institute
Bethesda, Maryland

Jane Sarnoff
Medical/Pharmaceutical Writer
New York, New York

REVIEWERS

Sandy Daack-Hirsch, RN, BSN
Genetics Nurse Specialist
Department of Pediatrics
University of Iowa Hospitals and Clinics
Iowa City, Iowa

Cinthia Prows, MSN, RN
Clinical Nurse Specialist, Genetics
Children's Hospital Medical Center
Division of Human Genetics
Cincinnati, Ohio

Patricia Ringhand, BSN, RN
ISONG Education Chair
Children's Hospital Medical Center
Division of Human Genetics
Cincinnati, Ohio

Jane Sarnoff
Medical/Pharmaceutical Writer
New York, New York

Knowing others is intelligence;
knowing yourself is true wisdom.
Mastering others is strength;
mastering yourself is true power.

Tao Te Ching

The following activities were designed to help you begin to explore your own values, beliefs, and knowledge in preparation for participating in the care of patients and families with current and future genetic concerns. Self-awareness is a first step in providing quality nursing care that successfully integrates genetics into your practice.

Take time to complete these exercises in preparation for discussion with your colleagues and instructors. A duplicate activity has been included at the end of this book. When you have completed this book and the reassessment at the end of it, return to your responses and thoughts in this section and reflect on how they are the same or different.

Part 1: Demographics

1. How old are you? _____ years of age

2. What is your current marital status?

    _____ Married

    _____ Widowed/divorced/separated

    _____ Never married

3a. Do you have any children?

    _____ No (skip to question 4)

    _____ Yes

 b. How old is your youngest child? _____ years of age

 c. Have any of your children been born with a birth defect or other congenital condition?

    _____ No

    _____ Yes

4. Is there any family history of genetic conditions or diseases in your family?

_____ No

_____ Yes

_____ Not sure

5. Have you, or has anyone in your immediate family, ever received genetic counseling?

_____ No, no one

_____ Yes, respondent

_____ Yes, other family member

_____ Not sure

6. In which ethnic group do you classify yourself?

_____ Hispanic

_____ American Indian or Alaskan Native

_____ Asian or Pacific Islander

_____ African American

_____ Causasian

_____ Other (please specify: _____)

7a. Have you ever attended any courses in human genetics?

_____ No

_____ Yes (please specify: _____)

b. Did you consider this course material relevant to nursing?

_____ No

_____ Yes

8. How thoroughly was genetics covered in other courses required for your nursing-education programs?

_____ Not covered

_____ Inadequately covered

_____ Adequately covered

_____ Very thoroughly covered

9. What are your *main sources* of information about new health problems and health care practices?

    _____ Professional journals

    _____ Professional associations

    _____ Nurses

    _____ Doctors

    _____ Other health professionals

    _____ Seminars and conferences

    _____ Continuing-education courses

    _____ Newspapers and news magazines

    _____ Other (please specify: _____)

Part 2: Professional History and Goals

*Place yourself on the scale with an **X**, describing realistically your knowledge of human genetics (genetics nursing) or your application of human genetics.*

1. My experience in this area of practice has been

None	Minimal	1–2 years	2–5 years	5–10 years	more than 10 years

2. My working knowledge of the current literature in this area is that I have

None	Minimal understanding	A broad overview	Read nursing journals	Read beyond nursing litera-ture into other disciplines	A comprehensive understanding

3. My comfort level in this area of nursing is

Extremely low	Moderate	Totally at ease

4. The importance of incorporating genetic skills and knowledge into the work that I do is

Minimal	Some	A lot	Tremendous

5. By reading this book, I hope to learn as much as I can about

    a. _____   c. _____

    b. _____   d. _____

6. This book will help me in my work by _____

_____

7. My personal goal for reading this book is _____

_____

8. I would feel I had accomplished an important achievement if, when I have finished reading this text, I could _____

_____

_____

Part 3: Nursing Education in Genetics: Topics Review

The purpose of this section of the self-assessment is to help you to clarify your views on topics that are important to nursing education in genetics. This survey is meant to be completed just before reading the text and again after reading the text, to examine your views and assess whether and how they have changed.

1. Below is a list of topics that are included in the text; for each topic, please indicate (by circling the number) whether you consider it
    1   Very important
    2   Somewhat important
    3   Not very important
    4   Not at all important
    5   Not sure or don't know

    a. Advantages and disadvantages of participating in family studies of genetic conditions   1   2   3   4   5

    b. Who has access to genetic information   1   2   3   4   5

    c. Screening of the general public for genetic disorders   1   2   3   4   5

    d. Treatments for genetic disorders, including gene therapy   1   2   3   4   5
    e. Including children in decisions about their genetic testing   1   2   3   4   5

    f. Coping with a new genetic diagnosis in the family   1   2   3   4   5

g. Genetic disorders and health insurance   1 2 3 4 5

h. Genetic information and employment   1 2 3 4 5

i. Genetic testing and biotechnology companies   1 2 3 4 5

j. New genetic information and the legal system   1 2 3 4 5

k. How to stay informed about new developments in the
    Human Genome Project   1 2 3 4 5

l. The media and interpreting the results of genetic
    research   1 2 3 4 5

m. How society may be affected by the Human Genome
    Project   1 2 3 4 5

n. Health care reform and the Human Genome Project   1 2 3 4 5

o. Family and professional partnerships   1 2 3 4 5

p. Basic human genetic concepts   1 2 3 4 5

q. Single-gene inheritance   1 2 3 4 5

r. Components of a family history   1 2 3 4 5

s. Molecular genetic clinical applications   1 2 3 4 5

2. Are there issues or topics related to genetic education for nurses that are not cov-
    ered in this book but that you consider important?

    _____ No

    _____ Yes (please comment below)

    Comments: _____

    _____

    _____

Genetics in Clinical Practice

The Human Genome Project

There's always a way—if you're committed.

Anthony Robbins

OBJECTIVE

Relate the impact of the Human Genome Project to clinical practice.

RATIONALE

The role of nurses is evolving as new genetic knowledge, gained from the achievement of the Human Genome Project's goals, is integrated into nursing practice. Being knowledgeable about the Human Genome Project will help nurses to apply the latest genetic information to patient care. This effort will entail the development of personal and professional approaches to keeping current on the most recent discoveries.

APPLICATION ACTIVITIES

- Assist patients and other professionals to understand the future impact of the Human Genome Project research results.
- In collaboration with other health professionals, assist in the development of goals that complement those of the Human Genome Project.

GENE
CARE
LINK

For up-to-date information about progress within the Human Genome Project, go to http://www.jbpub. com/clinical -genetics.

INTRODUCTION

The Human Genome Project (HGP) is one of the most significant scientific research endeavors of the twentieth century and has contributed to major advances in our understanding of the role of genetics in health and illness. These scientific discoveries are leading to new ways to diagnose, manage, and treat genetic conditions. The impact on all clinical practices will be tremendous and therefore has made professional educational efforts in genetics necessary at this time. Chapter 1 provides an overview of the major goals and technologies of the HGP and a discussion of some of the anticipated influences on nursing practice.

The HGP is the result of a collaborative, coordinated research effort in human genetics following the discovery of deoxyribonucleic acid, or DNA (Sorenson & Cheuvront, 1993). Launched by Congress in 1989, the HGP is the first centrally coordinated scientific effort to share genetic discoveries, data, and methods. The desired outcome of this centralized project is to facilitate understanding of the structure, function, and outcome of hereditary instructions found within the human genome. It is hoped that through these discoveries, medicine will be able to offer diagnostic, preventive, and treatment options for the more than 4,000 genetic diseases (Lessick & Williams, 1994). The HGP is likely to transform the role of medical, nursing, and other health care professionals by expanding available options for health care. This change must be anticipated now if we are to keep pace with the rapid rate of gene discovery. Our professional organizations and educational institutions must begin to assist professionals, patients, and the public in grappling with the future impact of such discoveries (U.S. Department of Health and Human Services [USDHHS], 1995a).

HISTORICAL PERSPECTIVE

The first proposal to develop a dedicated genome project emerged from discussion at a workshop at the University of Southern California, Santa Cruz, in 1985. The idea did not come to fruition until many additional discussions had taken place (Cook-Deegan, 1991). There was concern that shifting of scarce funding resources to such a project would displace other important work. Numerous interactions focused on which academic or federal agency would take the lead in centralizing the pursuit of genetic science (Friedman, 1990). Three reports, prepared for congressional review, helped pave the way for establishing the goals and focus of the HGP. These reports included a Report on the Human Genome Initiative prepared by the Health and Environmental Research Advisory Council of the Department of Energy, or DOE (Subcommittee on the Human Genome, 1987), a report by the U.S. Congress's Office of Technology Assessment (1988), and a National Research Council report (1988). On the basis of these reports and after significant discussion, in fiscal year 1988, congressional appropriations provided joint funding to the National Institutes of Health (NIH) and the DOE for genome research (USDHHS & USDOE, 1990). The agencies developed a memorandum of understanding in 1988 that outlined working relationships and a 5-year plan. James Watson was designated to head the program at NIH, and the National Center for Human Genome Research was established in Bethesda, Maryland, in 1989 to lead the NIH effort in coordinating genome research across the United States and in foreign countries (USDHHS, 1992). The major agencies supporting genome research projects include NIH, DOE, the National Science Foundation, and the Howard Hughes Medical Institute (Cutter et al., 1992).

STRATEGIC PLAN

It is estimated that between 80,000 and 100,000 human genes exist that ultimately control every aspect of human life, from what we look like, to the health problems we might develop, to perhaps even the way we view the world. The leaders within the HGP, including the current director, Dr. Francis Collins, recognized early on that specific goals and direction needed to be established for the Project. An undertaking of such broad scope could not be attempted without significant coordination and orchestration. The Project has been instrumental in fostering a multidisciplinary group approach to research and in inspiring a positive attitude about the potential outcomes of the mission to decode the exact sequence of all 3 billion nucleotide bases that make up the human genome (USDHHS, 1993 & 1995b). The goals of the $3 billion, 15-year HGP are outlined in Table 1.1.

Genetic Mapping

A primary goal of the HGP is to make available descriptive maps of the order of genes and of the genetic distance between them on each chromosome (USDOE, 1992). These genome maps include both genetic linkage maps and physical maps (see Appendix D Factsheet "Genetic Mapping"). Genetic linkage maps are developed to assign locations of genes or specific DNA sequences (markers) to specific chromosomal locations on the basis of how frequently these markers are inherited together (linked) with other markers whose location is known. The linear sequence of genes along the chromosomes provides the fundamental basis for genetic linkage. Linkage maps depend on a property of cell division called crossing over, or recombination. A more detailed description of recombination and the scientific basis of genetics is found in Chapter 3.

The detection of cross-over events requires a means by which to distinguish between paternal and maternal copies (alleles) of a marker in any individual. The wide variety among DNA from different individuals, called *polymorphisms*, provides the necessary means. Among the first polymorphisms to be recognized were the restriction fragment-length polymorphisms (RFLPs). These polymorphisms are caused by differences in DNA sequences that are recognized by *restriction endonucleases,* enzymes that recognize a specific DNA sequence and cut the DNA at that point. If a specific DNA sequence is present, the endonuclease will cut the DNA; if the sequence is absent, the DNA will not be cut.

TABLE 1.1

Human Genome Project Five-Year Plan (1994–1998)

FOCUS AREA	GOAL
Genetic mapping	Develop new markers and technology that facilitate rapid genotyping. Complete the 2–5-centimorgan map.
Physical mapping	Complete a sequence-tagged-site (STS)–based map of the human genome.
DNA sequencing	Develop efficient approaches and technological capability to sequence 50 DNA megabases per year.
Gene identification	Incorporate known genes onto physical maps to facilitate scientists studying the role of the gene in disease.
Technology development	Develop automation and robotics to improve efficiency, cost, and effectiveness of genome research.
Model organisms	Complete genetic maps of other species to facilitate modeling and understanding of the human genome.
Informatics	Develop efficient data management, analysis, and distribution access.
Ethical, legal, and social implications	Focus on issues that require policy discussions and recommendations, such as privacy of genetic information available from genetic testing.
Training	Encourage development of scientists and interdisciplinary genome researchers.
Technology transfer	Encourage transfer of discoveries into and out of genome research laboratories and centers.
Outreach	Share all information and materials within six months of development.

Source: Adapted from Collins & Galas, 1993.

In the 1980s, a technique called the *Southern blot* was developed to detect the presence or absence of restriction sites in the DNA of individuals. A person who is *heterozygous* for the presence of a particular restriction enzyme recognition site would have two or more DNA fragments resulting from the cleavage of his or her DNA by that enzyme. A person who is *homozygous* for the absence of the site would have only one, longer fragment, whereas a person homozygous for the presence of the site would have only one, shorter fragment (Figure 1.1).

RFLPs provided a tremendous amount of information to gene mappers in the 1980s and early 1990s. They are limited in their utility, however, because they are based on a two-allele system: on a single chromosome the site is either present or absent; there are no other options. Consequently, the use of RFLPs for gene mapping was limited. The recognition that there are other types of polymorphic systems within the human genome that could provide more information has advanced the process of gene mapping. The new type of polymorphisms involves variable numbers of tandem repeats or variable numbers of dinucleotide repeats. These tandem repeats may be as long as 15 to 20 nucleotides or as short as 2. This approach uses multiple alleles and provides information about genetic linkage in 80 to 90 percent of families studied. Variable numbers of tandem repeats and variable numbers of dinucleotide repeats have been extremely useful in the process of developing a dense genetic linkage map of the human genome.

Another remarkable technique that has rapidly advanced the field of molecular genetics and gene mapping is the polymerase chain reaction, or PCR. In this process, millions of copies of a selected region of DNA can be generated by an enzymatic reaction in only a few hours. In many cases, the PCR reaction can be substituted for much more complex and time-consuming procedures that take many days.

More precise genetic maps will enable investigators to localize genes that cause rare hereditary disorders and genes that contribute to more common conditions such as cancer, diabetes, and heart disease. Localization of these genes on a genetic map is the first step in identifying and characterizing them.

FIGURE 1.1

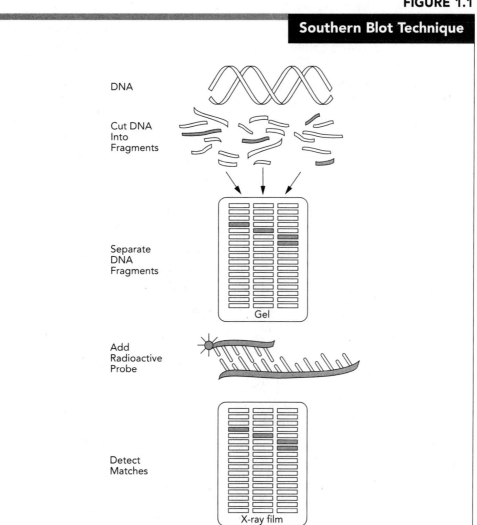

Southern Blot Technique

DNA

Cut DNA
Into
Fragments

Separate
DNA
Fragments

Gel

Add
Radioactive
Probe

Detect
Matches

X-ray film

Enzymes called restriction enzymes *recognize a specific DNA sequence and cut the DNA at that point. DNA fragments are separated. To find a target gene mutation in a sample of DNA, scientists then use a DNA probe, a length of single-stranded DNA that matches part of the gene and is linked to a radioactive atom. The single-stranded probe seeks and binds to the gene. Radioactive signals from the probe are then made visible on x-ray film, showing where the probe and gene are matched. (Adapted from National Institutes of Health and National Cancer Institute,* Understanding Gene Testing *[NIH pub. no. 96-3905]. Washington, DC: US Department of Health and Human Services, 1995.)*

Physical Mapping

Whereas genetic-linkage maps are used to assign genes to a specific area on a chromosome, physical maps are used to get a close-up view of a gene's precise location. A physical map is constructed by breaking a chromosome or chromosomes into smaller pieces of DNA. The physical-mapping goal is to complete a sequence-tagged sites (STS) map (Collins & Galas; 1993) that covers all the human chromosomes within a resolution of approximately 100,000 base pairs apart (see Appendix D Factsheet "Physical Mapping").

The development of a physical map depends on the duplication, or cloning, of fragments of DNA along a region of interest of a specific chromosome. The fragments of DNA to be replicated or cloned may be extremely large or relatively small. The size of the fragments one must study for any given region determines the type of "library," or collection of DNA fragments, one will use for that purpose. Genome scientists from all over the world have generated human libraries in a number of different carrier molecules called *vectors*. These libraries are now available to scientists engaged in the genetic and physical mapping of the genome and in the sequencing effort.

Positional Cloning

The process of positional cloning was developed by Francis Collins and other genome scientists to enable one to move from a linked genetic marker for a disease location to the actual gene causing the condition. Positional cloning depends on the chromosomal location of the gene only. This differs from traditional strategies of disease-gene identification, which required that one begin with the protein, determine the amino acid sequence, and use that information to isolate the gene. The process of positional cloning identified genes for several of the more common Mendelian disorders, including cystic fibrosis, neurofibromatosis type 1, and ataxia telangiectasia. Appendix D Factsheet "Positional Cloning" illustrates the process of positional cloning used to find the ataxia telangiectasia gene.

The process of positional cloning was also used to identify the gene for Huntington's disease, which in 1984 was linked to markers on human chromosome 4. This process took 10 years. The development of a much denser physical and genetic map in the years since then has made this process less arduous.

Positional Candidate Analysis

Positional candidate analysis is another molecular technique used to identify genes responsible for genetic disorders. A gene becomes a positional candidate by virtue of its location on the genetic map. An example of positional candidate analysis can be found in the discovery of the location of the gene for the human dwarfing condition known as *achondroplasia*. In 1994, the location for this gene was linked to markers in the region of human chromosome 4 where the gene for Huntington's disease was located. Within six months, the gene for achondroplasia had been found, due in no small part to the extensive efforts generated and the genes found in that region during the search for the location of the gene for Huntington's disease.

A gene called *fibroblast growth factor receptor 3* (FGFR3) had been found in the search for the gene for Huntington's disease. The gene was known to be expressed in cartilage. Because it sat within the candidate region determined by the linkage studies done to locate the gene for achondroplasia, FGFR3 became a positional candidate for the achondroplasia gene. Scientists searched for gene mutations in FGFR3 in the DNA of people who had achondroplasia and found gene mutations in all those studied.

As human genes and expressed gene sequences are placed on the genetic linkage and physical maps with ever-increasing frequency, the positional candidate approach to gene identification will become an increasingly viable option.

DNA Sequencing

The understanding of the structure and, ultimately, the function of the human genome requires determination of the linear sequence of all DNA base pairs (adenine, cytosine, guanine, and thymine) along the chromosome. When a genetic linkage map indicates that a gene lies in a particular region, clones of interest can be retrieved from the library for sequencing. Gene sequencing is the process by which the order of each base in the gene of interest is identified (see Appendix D Factsheet "DNA Sequencing"). This challenge of attaining the complete human DNA sequence is one of the major reasons for the emphasis on technology development within the Human Genome Project.

Model Organisms

The simpler systems of nonhuman genomes provide models for testing and developing data and procedures that can be used in studying the more complex human genome. Types of bacteria, yeast, the fruit fly, the roundworm, and the laboratory mouse have been used to develop physical maps and sequence genes. Comparison of their genomes with that of the human have already yielded important information related to human health and disease (Hoffman, 1994).

Technology Development

To date, the first two goals of the HGP—completing genetic and physical mapping of the human genome—have been attained. The major thrust of the Project at this point is to complete the sequencing of all genes. Progress in this area depends on the continuing support of technology development. Availability of methods that improve on the efficiency of DNA sequencing will facilitate the expediency and usefulness of this data in the future. Application of the information derived from the HGP in a variety of sites such as clinical and research laboratories and physicians' offices, will be increasingly desired, which will necessitate the development of new, cost-effective methodology.

Informatics

The amount of information to be generated through the HGP is mind boggling. The human genome alone contains approximately 3 billion base pairs, and the nature of the sequencing effort is such that much of the genome will be sequenced more than once before the project is complete. This does not even begin to consider the model organism genomes that are also being mapped and sequenced, and the huge array of genetic and physical mapping data, all of which must be stored. It also is desirable to make these data available to investigators worldwide in a timely manner. Computers offer the tools necessary to collect, organize, interpret, and disseminate such information, but software and database development and maintenance are needed. Patient-related information systems that ensure confidentiality are needed to facilitate family studies and future clinical genetic trials. Computer technology can also be used to provide training, access to resources and options for care, and other supportive information (Hubbard et al., 1995).

Ethical, Legal, and Social Implications

The scientific progress of the HGP will have a significant impact on future societal and individual decision options. Early in the evolution of the HGP, policymakers decided to dedicate a fixed percentage of funds for the Project to consideration of its societal implications. Hence, a major goal for the Project is to anticipate issues, solicit public and professional opinion, and develop guidelines or policy to address ethical, legal, and social concerns resulting from research on the human genome (Juengst, 1994). This is the so-called ELSI arm of the HGP. Education about these sensitive topics is critical to the 5-year plan's success. (See Chapter 9 for a more detailed discussion.)

Training

The success of the HGP will result in an increased demand for individuals trained in genetics. It is crucial that laboratory physicians, primary care providers, other health care professionals, and other individuals affected by these discoveries be able to communicate. The challenge is development of the skills, knowledge, and language necessary for understanding and communicating the impact of the available information. There is a significant need for multidisciplinary collaboration among the professional schools and societies, educational researchers, foundations, and government agencies involved in the HGP.

Technology Transfer and Outreach

The HGP will not be a true success unless its fruits are applied to the overall benefit of society. The collaboration of academicians, the private sector, and government agencies will facilitate enhanced transfer of technology. Because products should be made available to the community in a timely manner, sharing of information and materials within 6 months of their development has been stressed by the HGP. This sharing among researchers and organizations has permitted the rapid synergistic advances of the HGP (Hoffman, 1994). Critics have accused researchers of overplaying the potential advantages of the Project. It is crucial that researchers and companies present a balanced and realistic position to the public, working in collaboration with an educated media, so that expectations for immediate and effective applications are not unrealistically elevated.

CONCLUSION

The delineation of the human genome will revolutionize biological understanding of disease and future health care options. The organized and collaborative approach to gene identification represented by the HGP paves the way to understanding mechanisms for disease, to assessing the impact of environment on health, to developing a preventive health care focus, and potentially to developing innovative and more effective therapies. The discoveries made by the Project will also create ethical, legal, and social questions that are a significant focus of the HGP's five-year goals. This scientific initiative will have a major impact on all health care providers. The role for its discoveries in medicine's future is complex and challenging.

SUMMARY POINTS

The Human Genome Project (HGP)

- Collaborative, coordinated research effort
- To facilitate genetic discoveries
- Begun by government in 1988
- Likely to transform health care
- Joint project of NIH and DOE

National Center for Human Genome Research (NCHGR)

- Established in 1989
- Purpose is to head the role of NIH in coordinating genetic research
- Located in Bethesda, MD
- Director: Dr. Francis Collins

Strategic plan

- Joint NIH and DOE plan
- Sets the goals and direction of the HGP
- Multidisciplinary group approach
- Ultimate goal: provide tools that identify molecular and biochemical defects that result in disease

QUESTIONS FOR CRITICAL THINKING

1. What would be the advantages of having a centralized genetics research effort? disadvantages?

2. Which of the items identified within the 5-year plan will influence your practice the most? Why?

REFERENCES

Collins, F., and Galas, D. (1993). A new five-year plan for the US Human Genome Project. *Science* 262(5130):43–46.

Cook-Degan, R. (1991). The genesis of the Human Genome Project. *Mol. Genet. Med.* 1(1):1–75.

Cutter, M.A., Drexler, E., McCullough, L., et al. (1992). *Mapping and sequencing the human genome: science, ethics, and public policy.* Colorado: BSCS and the American Medical Association.

Friedman, T. (1990). The human genome project: some implications of extensive reverse genetic medicine. *Am. J. Hum. Genet.* 46(3):407–414.

Hoffman, E. (1994). The evolving genome project: current and future impact. *Am. J. Hum. Genet.* 54(1):129–136.

Hubbard, S., Martin, N., and Thurn, A. (1995). NCI's Cancer Information Systems: bringing medical knowledge to clinicians. *Oncology* 9(4):302–314.

Juengst, E. (1994). Human genome research and the public interest: progress notes from an American science policy experiment. *Am. J. Hum. Genet.* 54(1):121–128.

Lessick, M., and Williams, J. (1994). The human genome project: implications for nursing. *Med. Surg. Nursing* 3(1):49–58.

National Research Council (1988). *Mapping and sequencing the human genome.* Washington, DC: National Academy Press.

Subcommittee on the Human Genome (1987). *Report on the Human Genome Initiative: Health and Environmental Advisory Committee for the Office of Health and Environmental Research.* Washington, DC: DOE.

Sorenson, J., and Cheuvront, B. (1993). The Human Genome Project and health behavior and health education research. *Health Educ. Res.* 8(4):589–593.

US Congress, Office of Technology Assessment (1988). *Mapping our genes—genome projects: how big? how fast?* (rep. no. OTA-BA-373). Washington, DC: Government Printing Office. (Reprinted by Johns Hopkins University Press.)

US Department of Energy (1992). *Primer on molecular genetics.* Washington, DC: Department of Energy.

US Department of Health and Human Services (1992) *The Human Genome Project: new tools for tomorrow's health research* (rep. no. NIH 92-3190). Washington, DC: National Institutes of Health.

US Department of Health and Human Services (1993). *National Center for Human Genome Research progress report, fiscal years 1992 and 1993* (rep. no. NIH 93-3550). Washington, DC: National Institutes of Health.

US Department of Health and Human Services (1995a). *The Human Genome Project: from maps to medicine* (rep. no. NIH 95-3897). Washington, DC: National Institutes of Health.

US Department of Health and Human Services (1995b). *The Human Genome Project: progress report, fiscal years 1993–1994.* Washington, DC: National Institutes of Health.

US Department of Health and Human Services and US Department of Energy (1990). *Understanding our genetic inheritance: the first five years, FY 1991–1995* (rep. no. DOE/ER-0452P). Springfield, VA: National Technical Information Service.

Genetics and Health Care

The great end of life is not knowledge but action.

Thomas Henry Huxley

OBJECTIVE

Identify relevance of genetics to nursing practice.

RATIONALE

Gaps currently exist in the ideal and actual educational preparation of nurses for integrating genetics into patient care. Knowledge of the history of genetics in health care will prepare nurses to incorporate scientific and clinical concepts into their practice.

APPLICATION ACTIVITIES

- Consider roles that nurses might play in providing care to patients who have genetic concerns.
- Participate in the design of health care resources necessary to meet patient needs related to genetic health care.

GENE CARE LINK

For more information about the role of the nurse in genetics, go to http://www.jbpub. com/clinical-genetics.

INTRODUCTION

G*enes ultimately control every process that the human body performs. Our understanding of how genes function has been advanced by emerging methodologies in the laboratory (McKusick, 1996). Services, resources, and education in genetics have evolved into a genetics medical specialty in response to the need for the health care system to use this information for diagnosis, treatment, and prevention of genetic disorders.*

MEDICINE

The American Society of Human Genetics (ASHG), a professional organization founded in 1949, recommended in 1990 the establishment of the American College of Medical Genetics, which provides certification boards for geneticists with PhD and MD degrees. The ASHG is now the primary professional organization for human geneticists in North America, consisting of more than 5,800 members, and publishes the *American Journal of Human Genetics.* The society also sponsors congresses at which both laboratory and clinical advances are presented (McKusick, 1993). The first congress was held in 1956 and, within the next five years (1956–1961), aberrations of the chromosomes in Down syndrome, Klinefelter syndrome, and Turner syndrome were defined (McKusick, 1992). By 1966, the first edition of *Mendelian Inheritance in Man,* a catalogue of human genes and genetic disorders, had been published. The ability to use amniocentesis for prenatal diagnosis of genetic disease became possible in the 1970s. Over the next 20 years (1966–1986), many human genes were cloned, the location of the genes for other genetic diseases were discovered (mapped), and new techniques evolved, making genetics even more applicable in a clinical setting. The progress made from the late 1980s to the present has taken genetics from primarily an academic setting to a service more widely available within the community (Chen & Wertelecki, 1994). During the last 25 years (1956–1991), human genetics has become "medicalized, subspecialized, professionalized, molecularized, consumerized, and commercialized" (McKusick, 1992, p. 667). Soon it will become publicized as that common thread of knowledge that all health care professionals must include among their skills when offering health care services.

GENETICS SERVICES

Traditionally, genetics services have been provided by specially trained health professionals in academic medical settings, usually with a strong research and service emphasis (Andrews et al., 1994). Federal funding for such services has been available since 1978 to ensure access to comprehensive genetics services for all populations (Forsman, 1994). Implementation was made possible through the establishment of statewide systems of such services. Satellite genetics clinics have also been established in many states in primary and secondary sites, such as state health departments or community hospitals, to help facilitate accessibility. Expansion of genetics services into the community has served to engage many more primary care providers and other health professionals in genetic referral, counseling, evaluation, and follow-up services (Forsman, 1994).

Beginning in 1981, the regional genetics networks were established to continue support to genetics services. These regional networks provide support in the following areas: (1) networking among multidisciplinary providers; (2) planning and implementation of continuing education programs; (3) provision of quality assurance in laboratory programs; and (4) data collection. Nine regional genetics networks in the United States serve all 50 states, Puerto Rico, and the District of Columbia. The regional genetics networks continue to provide opportunities for genetics professionals, consumers, and other interested parties to engage in multidisciplinary activities that will advance genetics evaluation and services (Forsman, 1994).

Health professionals who provide specialized genetics services within each region usually include physicians, Ph.D. clinical geneticists, genetics counselors, nurse specialists, and social workers. As a result of the changing medical genetics environment, clinical genetics sites have begun to vary. Most genetics health professionals continue to practice in tertiary care or large community hospitals, where they may be involved in providing genetics counseling and evaluation for a regional clinic or a specialty genetics clinic such as a cystic fibrosis or hemophilia clinic. A smaller number of genetics professionals are in private (solo) practice, are employed by large health maintenance organizations, or are based in state departments of health or federal agencies. An increasing number are being employed by commercial agencies (Andrews et al., 1994).

In each of these locations, genetics professionals provide specified genetics services to primary care providers and their patients. Medical geneticists, genetic counselors, and genetic nurse specialists are available for (1) obtaining and interpreting complex family history information; (2) providing detailed explanations of genetic testing; (3) providing further information regarding protection of patient privacy; and (4) interpreting complicated genetics test results. Provision of continuing educational opportunities is another important service provided by most genetics health care professionals. Genetics centers, health professionals, and organizations can be identified through the regional genetics networks, the Council of Regional Genetics Networks (CORN), or individual genetics specialty organizations (see Appendix C).

Nurses may also call on genetic specialists for information about genetic conditions or available genetic testing; referral of an individual or family for more detailed genetic evaluation and counseling; coordination of services for families with genetic conditions; and collaboration in managing care of individuals with genetic conditions. Such collaborative activities will support primary care providers in ensuring that all their patients and families receive the most benefit from advancing genetic technologies (Williams & Lea, 1995).

The need for genetic counselors developed in the 1970s. A greater capability for prenatal diagnosis of genetic diseases brought families to pediatric, obstetric, prenatal diagnostic, and community health settings for case finding and counseling (Forsman, 1994). Health care team members in these settings had a Master's degree in genetics and were initially accredited through the ASHG. Genetics counselors now have their own professional society, the American Board of Genetic Counseling, which offers credentialing to its more than 1,000 members, and they are instrumental in identifying and counseling patients and families at risk for disorders that may have a significant genetic component (American Board of Genetic Counseling, 1995). Evaluation of health history, including family pedigree, lifestyle, and work exposures, and assessment of laboratory tests are examples of the counselor's role. In addition, presentation and discussion of prevention and intervention options are reviewed by the counselor with patient and family members, to assist them in making well-informed health and reproductive decisions related to genetic disorders.

The International Society of Nurses in Genetics (ISONG), a nursing specialty organization dedicated to fostering the scientific and professional growth of nurses in the field of genetics, was established to provide a forum for education and support for nurses providing genetic health care (Mathews, 1992). Incorporated in 1987, the ISONG promotes the integration of the nursing process into the delivery of genetic health care; encourages the incorporation of the principles of human genetics into all levels of nursing education; promotes the development of standards of practice for nurses in human genetics; and supports advancement of nursing research in human genetics.

ISONG members are present throughout the United States, Canada, Europe, and Taiwan. Membership is diverse and reflects the broad needs of those who have genetic problems. Current members include case managers, administrators, coordinators of public and private programs, educators in the field of nursing and genetics, and genetic counselors and researchers. Genetics nurse specialists are another important resource for nurses in all practice settings, providing genetics health education, counseling, advocacy, and support in collaboration with other health professionals.

ECONOMICS

Medical and clinical genetics services that currently are offered are preventive and are therefore often not covered by insurance. Many of the genetic discoveries, diagnostic capabilities, and resources for genetics services are still located in research or academic settings, thus requiring different types of financial support, such as grants or federal funding. Nonetheless, the genetic services provided are often time-consuming and labor-intensive and may not always be covered through state, federal, or private funds. The initial diagnostic laboratories were established in medical schools, where patients and families could have access to testing for limited or reduced fees. Now, numerous commercial DNA diagnostic services have successfully emerged (Bernhardt & Pyeritz, 1992), but as a result of this shift from academic-based to commercial genetics services, patients and families are more often faced with high fees for genetic testing and with concerns regarding insurance coverage. As diagnostic capabilities expand to the general population, significant attention will have to be paid to economic issues related to health care and the coverage of costs for available genetics services.

TREATMENT

Treatment options thus far for genetic diseases have been limited to include preventive activities such as prenatal diagnosis and symptomatic intervention. Options in the prenatal setting include using test results to determine whether a parent or fetus carries a dysfunctional hereditary trait. Decisions regarding ways in which to apply this information are often difficult. Families have required educational interventions to help them understand the clinical issues and have needed support in making decisions based on the information at hand. For example, instructions on ways to obtain follow-up support for making decisions regarding genetic information, prevention, and treatment options are crucial for a family in crisis. Understanding the risks and benefits of genetic tests and the significance of test results are critical elements of anticipatory guidance for persons or families in the decision-making process.

The discovery of genes is occurring rapidly, making diagnostic capabilities possible even before practical application of this knowledge is known or treatment outcomes are available. Genetic testing to identify individuals at risk for development of diseases such as breast or colon cancer is usually offered in research settings, although commercial availability is becoming an option. Current gene-therapy interventions are being explored in research settings as well; however, gene therapy for genetic disorders will likely be more widely used as clinical-trials outcomes become available (Carroll-Johnson, 1995).

EDUCATION

Nurses comprise the largest category of providers of care for patients and families, whether in an academic, community, or research setting; 2.2 million nurses are licensed to practice in the United States, and 1,853,024 registered nurses are currently employed (American Nurses Association, 1995). Formal genetic educational preparation of nurses has been lacking, although a recommendation to include this subject in general curricula was made in 1962 by Brantyl & Esslinger. Thomson (1993) reports that of the 200 advanced educational programs for nurses (Master's or Ph.D. programs), only three currently offer a genetics option at the graduate level. A comprehensive survey of 1,403 basic nursing programs in 1981 by Monsen (1984) illustrated that nearly half of the baccalaureate- and associate-degree nursing programs represented in the responses included five or fewer instructional hours in basic genetics concepts (mean 6.9 hours, range 0–173). In this same survey, nurses identified a need for improved instructional materials and aids to supplement knowledge and teaching of this content. Barriers to increasing genetics content in nursing curricula have included lack of appreciation for practical uses for this information, insufficient space in the curriculum, lack of faculty expertise, and funds inadequate to specialize in genetics (Kenner & Berling, 1990). Although many authors conclude that the field of genetics is dynamic and rapidly expanding and has the potential for substantially affecting the role of nurses, gaps in the ideal and the actual educational preparation of nurses currently exist (Monsen, 1992; Thomson, 1993).

A study, funded by the Human Genome Project of the National Institutes of Health, was recently implemented by the American Nurses Association (Scanlon & Fibison, 1995) to assess ways in which nurses in the United States are currently managing genetic information. Use of the survey outcomes has resulted in curriculum design and continuing education program development that will provide a focal point from which the knowledge gaps within the profession of nursing can be addressed.

The survey of 1,000 nurses examined attitudes and experiences with genetic testing, counseling, and referral. Only 9 percent of the respondents had taken a genetics course during their basic education. Almost 70 percent thought a course in human genetics should be a required course for a nursing degree. Most nurses did not feel knowledgeable in genetics (68 percent), although the majority had seen clients who had requested genetic information or had been referred for genetic counseling, screening, or testing. Respondents addressed attitudinal and ethical-dilemma case studies divergently, perhaps reflecting general societal responses. This finding indicates that it will be necessary to provide education related to the use of general ethical principles in evaluating such sensitive issues. The nurses did recognize the need for increased education, research funding, and safeguarding of genetic information. They expressed a desire for increased education in genetics for both nurses and the public. The study also reported that only 17 percent of the respondents had heard about the Human Genome Project. Nearly one-fourth of those responding believe that discoveries made by this scientific endeavor will *not* be relevant to nursing practice. Further educational initiatives will have to include information about the Human Genome Project, because it has the potential to change the role of nursing substantially (see Chapter 1).

CONCLUSION

Anticipation of a major shift in how health and disease is conceptualized, diagnosed, and managed as a result of technological development within genetics is the first step toward adequate preparation of nurses to incorporate scientific and clinical concepts into their practice. The shift from research and treatment to prevention and management of disease opens up new opportunities for public and professional education. Development of alternative care-delivery sites, providers, and methods will be necessary. Significant ethical, legal, and societal issues will have to be addressed. The health care resources within this country can be a major force in shaping genetics policy and practice. As we move into the twenty-first century, nurses in partnership with all health care providers can knowledgeably make decisions that influence and promote the delivery of quality, nondiscriminatory genetic health services.

SUMMARY POINTS

Genetic scientific discoveries will have several major effects.

- Influence available health care options
- Expand the role of health care professionals
- Challenge current values, beliefs, and knowledge

Preparation is needed by health care professionals and related resources to provide genetic health care.

- Physicians
- Genetic counselors
- Nurses
- Social workers
- Pastoral counselors

Preparation should involve review of what has been and consideration of what needs to be done in four basic areas.

- Professional societies
- Genetics services
- Economics and health coverage cost
- Educational preparation

QUESTIONS FOR CRITICAL THINKING

1. How will the future of health care be changed by the exploding discoveries within genetic science?

2. How might you prepare yourself professionally for applying clinical genetics in your practice?

3. Imagine yourself as part of a family with an extensive history of breast cancer. How would that history influence your interactions with patients with a similar background? Would you want to know whether you were at increased risk for developing breast cancer, and would you undergo genetic testing to determine that information?

REFERENCES

American Board of Genetic Counseling (1995). *Bulletin of information, description of examinations application form, 1996.* Bethesda, MD: American Board of Genetic Counseling, Inc.

American Nurses Association (1995). *Today's registered nurse: numbers and demographics.* Washington, DC: American Nurses Association.

Andrews, L., Fullarton, J., Holtzman, M., and Motulsky, A. (eds.) (1994). *Assessing genetic risks: implications for health and social policy.* Washington, DC: National Academy Press.

Bernhardt, B., and Pyeritz, R. (1992). The organization and delivery of clinical genetics services. *Pediatr. Clin. North Am.* 39(1):1–12.

Brantyl, V., and Esslinger, P. (1962). Genetics: implications for the nursing curriculum. *Nurs. Forum* 1(2):90–100.

Carroll-Johnson, R. (ed.) (1995). The genetic revolution. Promise and predicament for oncology nurses. *Oncol. Nurs. Forum* 22(2):3–36.

Chen, H., and Wertelecki, W. (1994). Genetic services in the United States. *Jpn. J. Hum. Genet.* 39(2):275–288.

Forsman, I. (1994). Evolution of the nursing role in genetics. *JOGNN* 23(6):481–486.

Kenner, C., and Berling, B. (1990). Nursing in genetics: current and emerging issues for practice and education. *J. Pediat. Nurs.* 5(6):370–374.

Mathews, A. (1992). International Society of Nurses in Genetics. *J. Pediatr. Nurs.* 7(1):3–4.

McKusick, V. (1966). *Mendelian inheritance in man.* Baltimore: Johns Hopkins University Press.

McKusick, V. (1992). Human genetics: the last 35 years, the present, and the future. *Am. J. Hum. Genet.* 50(4):663–670.

McKusick, V. (1993). Medical genetics: a 40-year perspective on the evolution of a medical specialty from a basic science. *J.A.M.A.* 270(19): 2351–2356.

McKusick, V. (1996). History of medical genetics. In D. Rimoin, J. Connor, and R. Pyeritz (eds.). *Emery-Rimoin principles and practice of medical genetics* (3rd ed.), pp. 1–30. New York: Churchill Livingstone.

Monsen, R. (1984). Genetics in basic nursing-program curricula: a national survey. *Matern. Child Nurs. J.* 13:177–185.

Monsen, R. (1992). Endpaper: a national agenda for nursing and genetics. *J. Pediatr. Nurs.* 7(1):63–64.

Scanlon, C., and Fibison, W. (1995). *Managing genetic information: implications for nursing practice.* Washington, DC: American Nurses Association.

Thomson, E. (1993). Reproductive genetic testing: implications for nursing education. *Fetal Diagn. Ther.* 8(suppl. 1):232–235.

Williams, J., and Lea, D. (1995). Applying new genetic technologies: assessment and ethical considerations. *Nurse Pract.* (20)7:16–26.

3

The Scientific Basis of Genetics

Stand before it and there is no beginning.
Follow it and there is no end.

Lao-Tzu

OBJECTIVE

Discuss concepts of basic human genetics and applications to nursing practice.

RATIONALE

New genetic technologies are allowing for better understanding of the genetic basis of health and disease. A sound understanding of basic human genetics will help nurses provide current information and care to patients and their families.

APPLICATION ACTIVITIES

- Educate patients and their families about genetic aspects of health and disease.
- In collaboration with other health-care providers, collect appropriate genetic family history and provide information about inheritance of genetic conditions and implications for other family members and for a broader community.

GENE CARE LINK

For excellent graphics and descriptions of molecular genetics, go to http://www. jbpub.com/ clinical-genetics.

INTRODUCTION

The essential characteristics of genes were first described by Gregor Mendel, an Austrian monk, in 1865 (Thompson et al., 1991). From his observations and analysis of the observable features of peas, he concluded that factors he called "particulate factors" were passed on unchanged from a parent plant to its offspring. These particulate factors are now known to be genes. During the last 20 years, new scientific technologies have allowed researchers to learn more about how genes function and how they influence health and disease. More than 4,000 genetic disorders are now known to be inherited in predictable patterns in families. The identification of genes related to such conditions as cancer and heart disease is leading to development of an increased number of genetic tests to diagnose a specific genetic condition or to provide information about a person's genetic predisposition to a condition influenced by both genes and environment. Genetic principles and information about their application to health and disease must be integrated into nursing education and practice. Nurses can then collect appropriate family history information, provide accurate and current genetic information, and contribute to the integration of this new information and technology into the daily lives of patients and their families.

This chapter provides a review of introductory information about the structure and function of genes, the genetic code, and inheritance of genetic disorders. It lays a foundation for subsequent chapters, which address the application of genetic principles and emerging genetic technologies to nursing practice. More detailed discussion and presentation of genetic principles can be found in the references at the end of the chapter.

GENES: THE UNITS OF HEREDITY

Each person's genetic makeup contains between 50,000 and 100,000 genes, and each gene is considered a unit of heredity. A *gene* is composed of a segment of deoxyribonucleic acid (DNA) that contains a specific set of instructions for making a protein needed by body cells for proper functioning. Genes control both the type of protein made and the rate at which proteins are produced.

The specific shape of the DNA molecule was identified by Watson and Crick in 1953 (Watson & Crick, 1953; Thompson et al., 1991) and has since been termed the *double helix.* The essential components of the DNA molecule are sugar-phosphate molecules (comparable to the railings on a spiral staircase) and pairs of nitrogenous bases (which serve as the backbone and are similar to the stairs that hold the railings of the spiral staircase). Each *nucleotide* contains a sugar (deoxyribose), a phosphate group, and one of four nitrogenous bases: adenine (A), cytosine (C), guanine (G), and thymine (T). DNA is composed of two paired strands, each of which is made up of a number of nucleotides. The strands are held together by hydrogen between pairs of bases (Figure 3.1).

The bases on each strand are always paired in the same way: adenine of one DNA strand is paired with thymine of the other strand, and guanine of one DNA strand is paired with cytosine of the other strand. The pairing is related to the formation of hydrogen bonds. Adenine and thymine form two hydrogen bonds; cytosine and guanine form three hydrogen bonds. The formation of these hydrogen bonds between the bases is referred to as *base pairing,* and the bases are said to be *complementary.*

The length of a DNA sequence of a gene is expressed in units of base pairs (bp) or thousands of base pairs, called *kilobase pairs* (kb). For example, the gene for Duchenne muscular dystrophy is 2,300 kilobases long, whereas the hemophilia gene is 186 kilobases long.

Twenty different *amino acids* are found in the human body, and these are the building blocks of proteins. The sequence of nucleotides within DNA determines the sequence of the amino acids in the gene product. Each amino acid is determined by a set of three

FIGURE 3.1

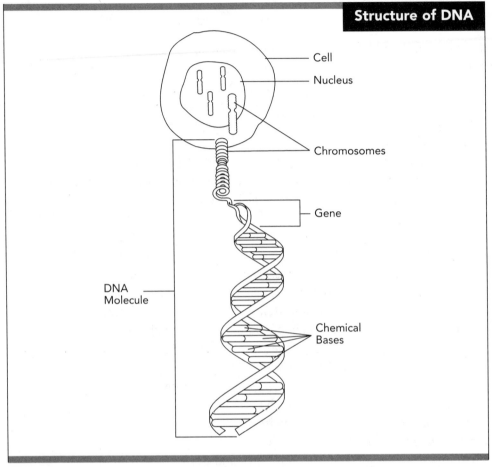

DNA, which carries the instructions that allow cells to make proteins, is made up of four chemical bases: adenine (A), guanine (G), cytosine (C), and thymine (T). Tightly coiled strands of DNA are packaged in units called chromosomes, *housed in the cell's nucleus. Working subunits of DNA are known as* genes. *(Adapted from National Institutes of Health and National Cancer Institute,* Understanding gene testing *[NIH pub. no. 96-3905]. Washington, DC: US Department of Health and Human Services: 1995.)*

bases called a *triplet.* Amino acids can be specified by more than one triplet, but each triplet makes only one amino acid. The amino acid phenylalanine, for example, may be specified by the triplets UUU, UUC, UUA, UUG (Levitan, 1988; Thompson et al., 1991). The relationship of specific triplets to specific amino acids is called the *genetic code.*

CHROMOSOMES: THE PACKAGES OF GENES

DNA is packaged in cellular structures called *chromosomes.* The genes are arranged in a linear order along each chromosome, with each gene having a specific chromosomal location, known as its *genetic locus.* Chromosomes are located in the nucleus of the cell and occur in pairs; of each pair, one chromosome is inherited from the mother and one from the father. Each human cell contains 46 chromosomes (23 pairs). The only exception is the germ cell (egg or sperm), which contains only one chromosome of each pair (23 chromosomes). The 22 pairs of chromosomes that are alike in males and females are called *autosomes:* The chromosomes of each pair are said to be homologous to each other. The twenty-third pair constitutes the *sex chromosomes:* females have two X chromosomes, and males have one X and one Y chromosome. A female inherits one X chromosome from each parent. A male inherits an X chromosome from his mother and a Y chromosome from his father (Figure 3.2).

Careful examination of DNA sequences from many individuals reveals that many sequences have multiple different versions in a population. These different versions, or sequence variations, are called *alleles.* Sequences that are found in many forms are said to be *polymorphic,* meaning that there are at least two common forms of a particular gene. When a person has identical alleles at a specific genetic locus, he or she is said to be *homozygous* at that locus. If, on the other hand, alleles are not the same, a person is said to be *heterozygous.*

CELL DIVISION

The human body grows and develops by a process of cell division. Two different kinds of cell division—meiosis and mitosis—contribute to this process.

Meiosis is the cell division process by which oocyte and sperm (gametes) are formed. This process is confined to reproductive cells.

FIGURE 3.2

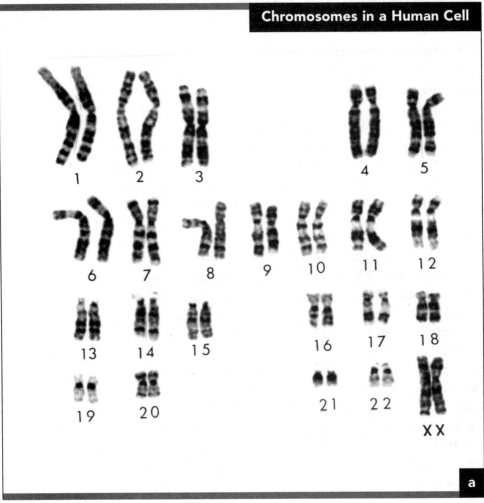

Chromosomes in a Human Cell

*Each human cell contains 23 pairs of chromosomes, which can be distinguished by size and unique banding patterns. Set **a** is from a female and contains two X chromosomes; set **b** is from a male and contains an X and a Y chromosome. (Reprinted with permission from David H. Ledbetter, PhD, FACMG.)*

During meiosis, a reduction in the number of chromosomes occurs through a series of complex mechanisms, so that the resulting gamete contains 23 chromosomes, or one-half the original number (Figure 3.3). Because human oocytes and sperm contain only 23 chromosomes, they are said to be *haploid,* as compared to somatic or

FIGURE 3.2

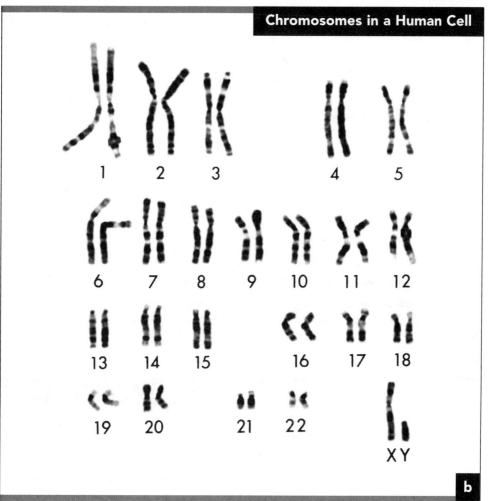

Chromosomes in a Human Cell

1   2   3   4   5

6   7   8   9   10   11   12

13   14   15   16   17   18

19   20   21   22

X Y

b

body cells, which have 23 *pairs* of chromosomes and are said to be *diploid.* The process of meiosis occurs in two stages of cell division. In the first stage (meiosis I), a phenomenon called *crossing over* may occur: Paired chromosomes come together and portions cross over, or exchange sections of genetic material. This recombination of genetic material inherited from the father and the mother leads to more diversity in the makeup of egg and sperm (Figure 3.4). Crossing over was discussed in more detail in Chapter 1.

FIGURE 3.3

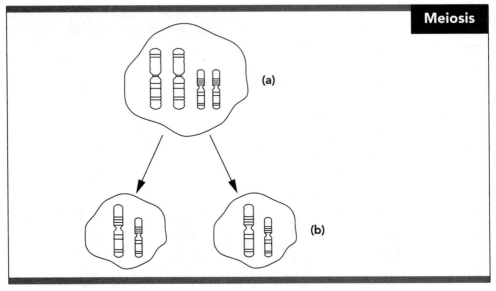

Meiosis is the cell-division process by which egg and sperm are formed. During meiosis, a reduction of the usual diploid number of chromosomes (a) occurs, resulting in an egg or sperm that contains one-half of the original number. An egg or a sperm containing 23 chromosomes is said to be haploid (b).

FIGURE 3.4

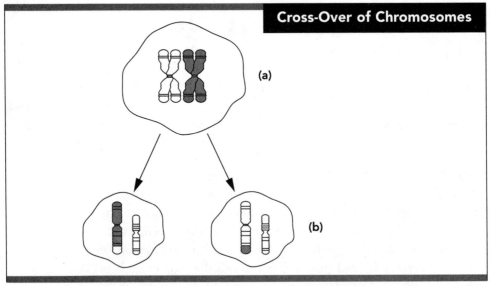

During meiosis, crossing over may occur: paired chromosomes come together (a), and portions cross over—that is, exchange sections of genetic material (b).

During either the first or second meiotic division, an accidental event called *nondisjunction* may occur: the paired chromosomes fail to separate, creating an egg or a sperm that contains either two copies or no copy of a chromosome. An embryo resulting from fertilization of a normal oocyte by a sperm that lacks one chromosome will have 22 pairs of chromosomes and a single chromosome that is missing its mate. This is called *monosomy*. Conceptions with monosomy of one of the 22 autosomes are usually lost spontaneously. When an egg or a sperm that contains two copies of a chromosome is fertilized, the result is called *trisomy* (Figure 3.5).

FIGURE 3.5

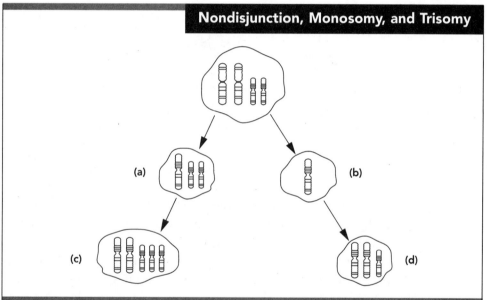

Nondisjunction, Monosomy, and Trisomy

(a) (b) (c) (d)

Nondisjunction may occur during meiosis. The paired chromosomes fail to separate, creating an egg or a sperm that contains either two copies of a chromosome (a) or no copy of a chromosome (b). These are unbalanced gametes. The result is a conception with three copies of a chromosome, called trisomy *(c), or only one copy of a chromosome, called* monosomy *(d).*

Mitosis is the process of cell division that determines cellular growth, differentiation, and repair. Mitosis occurs in all body cells except reproductive cells. During mitosis, the chromosomes of each cell duplicate, and the resulting two cells (daughter cells) each have the same number of chromosomes as does the parent cell. These

FIGURE 3.6

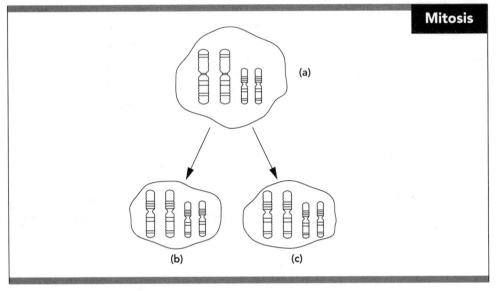

Mitosis involves a duplication of chromosomes in each cell. The parent cell (a) produces daughter cells (b), each of which has the identical number of chromosomes as has the parent cell.

cells are said to be diploid, because they have the usual two of each chromosome pair (Figure 3.6).

A number of different types of chemical stains are available to enable researchers to visually examine chromosomes under a microscope. The most commonly used of these is the Giemsa stain, which results in a banding pattern that allows unambiguous identification of each chromosome. The microscopic study of chromosomes is called *cytogenetics*. Cytogenetic analysis of a human cell, with banded chromosomes arrayed in order on a piece of paper, is called a *karyotype*. Sophisticated laboratory techniques developed in recent years have moved cytogenetics into the molecular age. The field of molecular genetics, combining molecular techniques with examination of chromosomes, is evolving rapidly.

DNA, RNA, PROTEIN: THE CENTRAL DOGMA OF GENETICS

The genetic information contained in DNA within a chromosome is located in the cell nucleus. The creation of a specific protein for which the DNA code is needed, however, occurs outside the cell nucleus in a part of the cell called the *cytoplasm*. When a normally functioning cell makes a protein, the process begins with the uncoiling of the DNA into its two strands, called the *coding,* or *sense, strand* and the *antisense strand.* Information from the coding strand is *copied* base by base from the DNA into a single-stranded nucleic acid, called messenger ribonucleic acid (mRNA), by an enzyme called *RNA polymerase.* After it is copied from the DNA, the mRNA

FIGURE 3.7

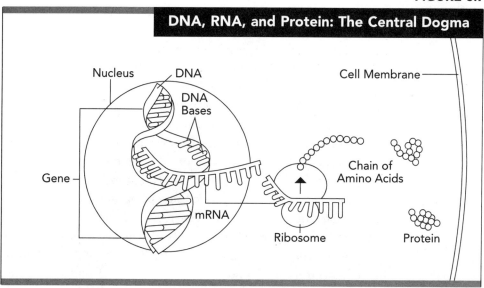

DNA, RNA, and Protein: The Central Dogma

Nucleus   DNA

DNA Bases

Cell Membrane

Gene

Chain of Amino Acids

mRNA

Ribosome

Protein

For a cell to make a protein, the information from a gene is copied base by base, from DNA into new strands of messenger RNA (mRNA). Then mRNA travels out of the nucleus into the cytoplasm, to cell organelles called ribosomes. There, mRNA directs the assembly of amino acids that fold into a completed protein molecule. (Adapted from National Institutes of Health and National Cancer Institute, Understanding gene testing *[NIH pub. no. 96-3905]. Washington, DC: US Department of Health and Human Services, 1995.)*

moves out of the cell into the cytoplasm and from there directs the assembly of amino acids to form a completed protein molecule. Protein synthesis takes place within the cell on structures called *ribosomes* (Figure 3.7).

The two main processes involved in making a new protein are *transcription* and *translation*. Transcription occurs when mRNA is copied from coding strands of DNA. Translation is the process by which the amino acids are assembled into proteins based on the triplet sequences of the genetic code. Transcribing DNA to RNA and translating RNA to a protein are the central processes of molecular genetics (Lewin, 1994).

GENE STRUCTURE AND FUNCTION

Genes are interspersed with coding and noncoding regions of DNA called *exons* and *introns,* respectively. Introns—portions of a gene located in the noncoding regions—contain sequences involved with gene regulation and RNA splicing. Introns are present (transcribed) within RNA in the nucleus of all cells but are deleted (spliced) from mRNA as it moves into the cytoplasm. Therefore, introns are not represented (translated) in the final protein product. Exons—portions of a gene located in the coding regions—contain all the genetic information translated into protein, as well as a number of sequences necessary for transcription and translation. In addition to coding sequences, proper expression of a gene to create a protein requires fixed starting and ending points. For example, the starting point for making mRNA is located in a position of the gene called the *promotor,* or *5' upstream flanking region.* An untranslated region at the end of the gene, called the *3',* or *downstream flanking region,* contains a signal for the end of mRNA. Transcription of DNA to RNA occurs from the 5' end systematically to the 3' end of the gene.

With the aid of a reading frame that divides the DNA sequence into triplets, one can analyze DNA sequences to determine actual and potential protein products. Because three bases comprise a triplet, a triplet can be read in three ways, depending on which nucleotide is selected as the first. Several computer programs are available that can analyze a DNA sequence according to all possible reading frames and deliver the translated protein sequence for all possible outcomes. These programs have made it much simpler to translate DNA sequences into protein sequences that can be analyzed for functional implications (Lewin, 1994).

GENE MUTATIONS

Gene mutations occur as a result of several different kinds of alterations in the DNA. The most common type of gene *mutation* involves a "misspelling" in the DNA sequence that results in the alteration of a single DNA base. In sickle-cell anemia, for example, one nucleotide in the gene coding for the beta-globin chain of hemoglobin is altered, producing hemoglobin S. Persons who have two copies of the gene mutation hemoglobin S have sickle-cell anemia, a disorder of hemoglobin structure and function leading to anemia and organ damage resulting from hypoxia.

Other gene mutations involve the addition (insertion) or loss (deletion) of one or more DNA bases or multiplication of a longer DNA segment (duplication). A relatively newly recognized type of gene mutation involves the expansion of the number of triplet repeat sequences within a gene. Some gene mutations, called *silent mutations,* have no significant effect on the structure or function of the protein product, whereas other gene mutations result in a partially or completely altered protein product. The effect of a gene mutation depends upon how a protein is altered and on the protein's importance to proper body functioning (Figure 3.8).

Gene mutations may be inherited from one or both parents, may occur as a new event at the time of gametogenesis (formation of the oocyte or sperm), or might be acquired in the course of mitosis any time after conception. An inherited gene mutation is present in DNA in all body cells and is passed on from generation to generation in reproductive cells. These inherited gene mutations are referred to as *germline mutations.* Germline mutations are copied and passed on to all daughter cells when body cells divide (Figure 3.9). The gene mutation that causes Huntington's disease is a germline mutation.

FIGURE 3.8

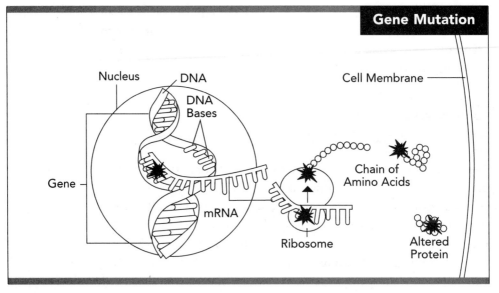

Gene Mutation

Nucleus   DNA

DNA Bases

Cell Membrane

Gene

mRNA

Chain of Amino Acids

Ribosome

Altered Protein

When a gene contains a mutation, the protein encoded by that gene will be abnormal. Some protein changes are insignificant, whereas others are disabling. (Reprinted from the National Institutes of Health and National Cancer Institute, Understanding gene testing *[NIH pub. no. 96-3905]. Washington, DC: US Department of Health and Human Services, 1995.)*

A new or spontaneous mutation occurs in an individual egg or sperm contributed to a conception and usually is not present in other family members. The person who carries the new gene mutation will have that mutation in all of his or her body cells and has an increased chance of passing on the gene mutation to offspring.

An acquired gene mutation, called a *somatic mutation,* is a change in DNA that develops after conception during the life span. Somatic mutations occur as a result of changes in the DNA found in the cells of the body but not in the germline. The gene alterations are passed on only to the descendants of that particular cell. Acquired germline mutations may not be present in body cells but may still be passed on to children. This situation is known as *germline mosaicism.*

Gene mutations occur regularly in body cells. Cells have the capacity, however, to recognize DNA changes and, in most instances, correct the alteration before it is passed on through cell division. Over time, cells' abilities to repair damage from gene mutations may diminish and lead to an accumulation of genetic alterations that

FIGURE 3.9

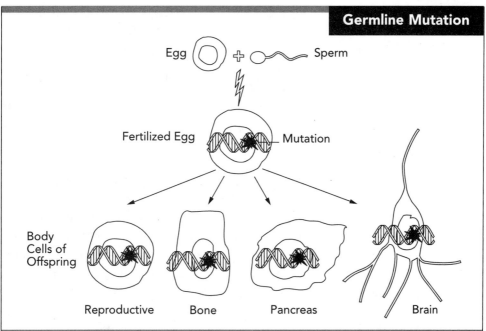

Hereditary mutations are carried in the DNA of the reproductive cells. When reproductive cells containing mutations combine to produce offspring, the mutation will be present in all of the offspring's body cells. (Reprinted from the National Institutes of Health and National Cancer Institute, Understanding gene testing [NIH pub. no. 96-3905]. Washington, DC: US Department of Health and Human Services, 1995.)

may ultimately cause disease. Accumulation of genetic alterations are often involved in cancer and may contribute also to other disorders of aging such as arthritis and dementia.

GENES: A CRITICAL COMPONENT OF HEALTH AND DISEASE

Regulation of the expression of the 50,000 to 100,000 genes located in the 46 human chromosomes occurs as a series of complex interrelationships within the cell. Gene structure, transcription, translation, and protein manufacturing processes are all involved. An alteration in any one of these may influence health and disease. More than 4,000 single-gene disorders have been identified and cataloged in Victor A. McKusick's *Mendelian Inheritance in Man* (1994). Single-gene disorders are caused by gene mutations that may be present on either one or both of the two chromosomes of a pair. In each of these situations, mutations at a single gene locus cause a genetic condition. Single-gene disorders are known also as *Mendelian disorders* because, like the characteristics observed by Mendel in his garden peas, they are inherited in families and occur with fixed proportion among generations.

Other health conditions with a genetic component, such as cancer and heart disease, are now known to occur as a result of multiple-gene mutations interacting with environmental influences. These are referred to as multifactorial genetic conditions, or complex traits. Still other genetic disorders are caused by a gross alteration in the number or structure of chromosomes. These conditions are called *chromosomal disorders*.

CLASSIFICATION OF HEREDITARY GENETIC DISORDERS

Most health conditions are believed to be the result of a combination of genetic and environmental influences and interactions. The relative contribution of the genetic component, however, may be large or small. The genetic contribution to a person's response to accidental injury may reside in that person's physiological capacity for tissue repair. In contrast, in a person who is born with Down syndrome or cystic fibrosis the genetic contribution is significant and can be modified by environmental or treatment options.

Each person has a specific genetic makeup or *genotype*. A person's *phenotype* is the observable manifestation of his or her genotype, whether it be the person's appearance, or a biochemical, molecular, or even psychological trait. Most phenotypes, even those with a very strong genetic component, are influenced by environmental factors. A person with sickle-cell anemia, for example, has a genotype of two alleles for the sickle-cell mutation at the beta-globin locus. This person's phenotype is the manifestation of symptoms of sickle-cell anemia, such as sickle-cell crises. The phenotype of sickle-cell disease can be modified by prophylactic antibiotics and by limited exposure of the patient to temperature extremes.

A single-gene (Mendelian) disorder is a condition that is caused by a gene mutation at a single location on one or both chromosomes of a pair. A person who has the same gene mutation on both chromosomes of a pair is said to be *homozygous* for that mutation. A person who has two different mutations at the same location is described as a *compound heterozygote*. An individual with one mutation and one "normal" allele at a single locus is said to be *heterozygous* for the mutation.

Single-gene disorders are characterized by their pattern of transmission. Three main types of inheritance for single-gene disorders have been identified: autosomal dominant, autosomal recessive, and X-linked disorders. It is important to note that the terms *dominant* and *recessive* are applied to the trait, disorder, or phenotype but not to the genes or alleles causing the observable characteristics.

The first step in establishing the pattern of inheritance is usually to obtain a family history and to summarize it in the form of a pedigree. Table 3.1 illustrates examples of standard symbols for pedigree construction.

Nursing assessment of individuals and families includes obtaining and recording a family history. Familiarity with the characteristics of single-gene inheritance and knowledge of pedigree construction is necessary so that recognizable patterns of inheritance can be identified and appropriate genetic information, further counseling, and testing can be offered (Williams, 1996).

Autosomal-Dominant Inheritance

Autosomal-dominant conditions follow a vertical pattern of inheritance in families and occur equally frequently in males and females of all ethnic groups (Figure 3.10a). A person who has an autosomal-dominant genetic condition has a single-gene mutation located on one chromosome of a pair. An affected individual generally has, with each reproductive event, a 50 percent chance for passing on the gene mutation or the normal form of the gene to offspring. Offspring

TABLE 3.1

Examples of Commonly Used, Standardized Pedigree Symbols, Definitions, and Abbreviations

Instructions:
— Key should contain all information relevant to interpretation of pedigree (e.g., define shading)
— For clinical (non-published) pedigrees, include:
  a) family names/initials, when appropriate
  b) name and title of person recording pedigree
  c) historian (person relaying family history information)
  d) date of intake/update
— Recommended order of information placed below symbol (below to lower right, if necessary):
  a) age/date of birth or age at death
  b) evaluation
  c) pedigree number (e.g., 1-1, 1-2, 1-3)

	Male	Female	Sex Unknown	Comments
1. Individual	b. 1925	30 y	4 mo	Assign gender by phenotype.
2. Affected individual				Key/legend used to define shading or other fill (e.g., hatches, dots, etc.).
				With ≥2 conditions, the individual's symbol should be partitioned accordingly, each segment shaded with a different fill and defined in legend.
3. Multiple individuals, number known	5	5	5	Number of siblings written inside symbol. (Affected individuals should not be grouped.)
4. Multiple individuals, number unknown	n	n	n	"n" used in place of "?" mark.
5a. Deceased individual	d. 35 y	d. 4 mo		Use of cross (†) may be confused with symbol for evaluated positive (+). If known, write "d." with age at death below symbol.
5b. Stillbirth (SB)	SB 28 wk	SB 30 wk	SB 34 wk	Birth of a dead child with gestational age noted.
6. Pregnancy (P)	P, LMP: 7/1/94	P, 20 wk	P	Gestational age and karyotype (if known) below symbol. Light shading can be used for affected and defined in key/legend.
7a. Proband	P↗	P↗	P↗	First affected family member coming to medical attention.
7b. Consultand	↗	↗		Individual(s) seeking genetic counseling/testing.

Instructions:
— Symbols are smaller than standard ones and individual's line is shorter. (Even if sex is known, triangles are preferred to a small square/circle; symbol may be mistaken for symbols 1, 2, and 5a/5b of Figure 1, particularly on hand drawn pedigrees.)
— If gender and gestational age known, write below symbol in that order.

	Male	Female	Sex Unknown	Comments
1. Spontaneous abortion (SAB)	male	female	ECT	If ectopic pregnancy, write ECT below symbol.
2. Affected SAB	male	female	16 wk	If gestational age known, write below symbol. Key/legend used to define shading.
3. Termination of pregnancy (TOP)	male	female		Other abbreviations (e.g., TAB, VTOP, Ab) not used for sake of consistency.
4. Affected TOP	male	female		Key/legend used to define shading.

Reprinted with permission from R. Bennett et al., Recommendations for standardized human pedigree nomenclature. J. Genet. Counsel. 4(4) (1995):267–279.

FIGURE 3.10

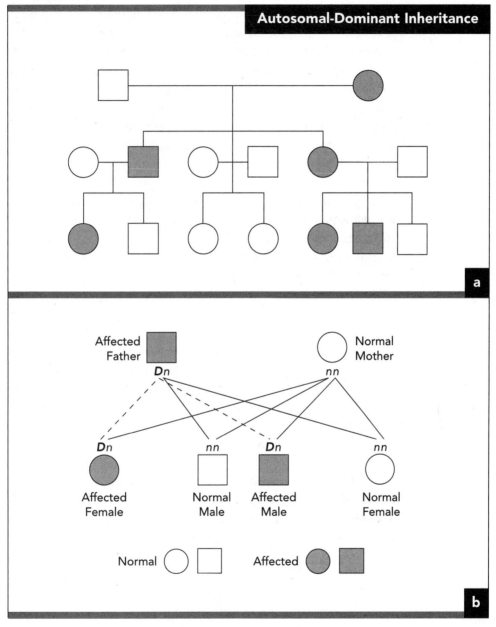

(a) *Autosomal-dominant family pedigree.* (b) *In autosomal-dominant inherited conditions, an affected person has a single, dominant (D) gene that causes the condition. With each pregnancy, he or she has a 50% chance of passing on the dominant gene for the condition and a 50% chance of passing on the normal (N) gene, which would* not *cause the condition to be present.*

who do not inherit the gene mutation that causes the dominant disorder will have neither the condition nor an increased chance for passing on the gene for the condition to their children (Figure 3.10b).

Advanced paternal age is one factor that has been associated with an increased chance for the occurrence of new autosomal-dominant conditions in a family as a result of new gene mutations during spermatogenesis. In such families, a condition known to be inherited as an autosomal-dominant disorder may appear for the first time in the offspring of unaffected parents. The condition in these patients is said to be sporadic, because it has not been seen previously in the family.

Other characteristics of autosomal-dominant inheritance include *variable expression* of the condition among affected family members and a phenomenon called *reduced penetrance*. With variable expression of a dominant-gene mutation, individuals who have the gene for a condition may show varying degress of severity. Some may have very mild symptoms, whereas others may have significant symptoms. Both genetic and environmental influences are believed to be related to variability in clinical presentation.

With reduced penetrance, a person may inherit an autosomal-dominant gene mutation that causes an autosomal-dominant disorder but may not demonstrate any physical or developmental characteristics of the condition. These so-called nonpenetrant individuals still have a 50 percent chance for passing on the gene mutation to their offspring even though they may not appear to have the condition seen in other members of the family. In counseling unaffected family members about their reproductive risks, the counselor must know the penetrance of the disorder. Features of autosomal-dominant inheritance and examples of such conditions are listed in Table 3.2.

Autosomal-Recessive Inheritance

Autosomal-recessive disorders are found with increased frequency among specific ethnic groups and tend to be seen more often in offspring of parents who are blood relatives (e.g., cousins). The pattern of inheritance observed in autosomal-recessive disorders is more horizontal than vertical, with multiple members of a single generation tending to be affected with the condition (Figure 3.11a).

TABLE 3.2

Autosomal-Dominant Inheritance

FEATURES	EXAMPLE CONDITIONS
Vertical transmission	Achondroplasia
Equal frequency of affected males and females	Colon cancer
	Hereditary breast or ovarian cancer
Advanced paternal age associated with sporadic cases	Huntington's disease
Variable expression	Marfan syndrome
Reduced penetrance (in some disorders)	

In autosomal-recessive inheritance, each parent, called a *carrier,* has a single-gene mutation on one chromosome of a pair and the normal form of the gene on the other. A carrier does not have the condition, but when two carrier parents reproduce, each reproductive event is characterized by a 25 percent chance of bearing a child who inherits the gene mutation from each parent and who will have the genetic condition. Carrier parents also have a 50 percent chance of giving birth to a child who inherits a single-gene mutation from one parent and is a carrier and a 25 percent chance of giving birth to a child who will inherit the normal form of the gene from each parent and will neither be affected nor be a carrier (Figure 3.11b). Features of autosomal-recessive inheritance and examples of such conditions are listed in Table 3.3.

X-Linked Inheritance

X-linked conditions are caused by mutations on the X chromosome. Because males have only a single X chromosome, they are said to be *hemizygous* for genes on the X chromosome. X-linked disorders may be recessive—that is, expressed only in males (Figure 3.12a). More rarely, an X-linked disorder may be dominant—that is, expressed in both males and females, even though the female carries a normal version of the gene on her other X chromosome.

FIGURE 3.11

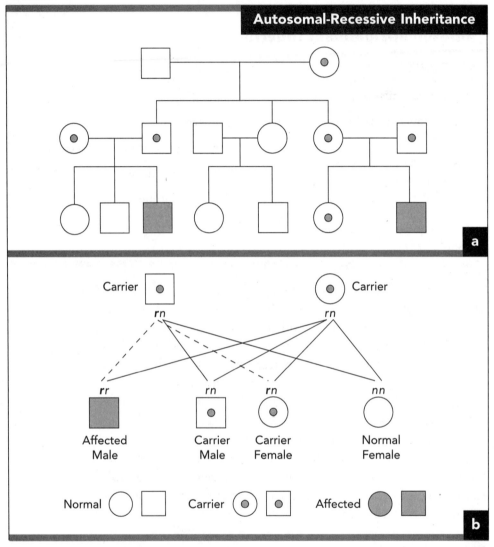

(a) Autosomal-recessive family pedigree. (b) In autosomal-recessive inherited conditions, each parent carries a gene for the condition (r) in one of a pair of chromosomes, and a normal gene (n) on the other chromosome of the pair. With each pregnancy, there is a 25% chance of having an affected child, a 50% chance of having children who would be carriers like the parents, and a 25% chance of having a child who inherits two normal genes and is neither affected nor a carrier.

TABLE 3.3

Autosomal-Recessive Inheritance

FEATURES	EXAMPLE CONDITIONS
Horizontal occurrence in a pedigree	Cystic fibrosis
Equal frequency of affected males and females	Hemochromatosis
	Phenylketonuria (PKU)
Associated with particular ethnic groups and consanguinity (blood-related parents)	Sickle-cell anemia
	Tay-Sachs disease

In X-linked-recessive inheritance, a female carrier has a 50 percent chance of passing on the gene mutation to a male child, who would be affected, or to a female child, who would be a carrier, like her mother. She also has a 50 percent chance of passing on the normal gene to a male, who would not be affected, and to a female, who would be neither affected nor a carrier (Figure 3.12b). In X-linked dominant disorders, an affected female has a 50 percent chance for passing on the gene mutation to a male or female offspring, who would be affected, and a 50 percent chance of passing on the normal gene to a male or female offspring, who would be neither affected nor a carrier. X-linked-dominant disorders tend to be much more severe in males than in females, and many of them are lethal in the hemizygous state.

X-linked genetic conditions are clinically variable in females because of a phenomenon called *X-inactivation* or Lyonization (after Mary Lyon, who first described the phenomenon; Lyon, 1962). Because female cells have two X chromosomes, one of the X chromosomes is inactivated, or "turned off," to maintain a relatively constant level of expression of X-linked genes in male and female cells. Some genes on the X chromosome escape the process, but most are expressed in only one copy in every cell. X-inactivation influences the degree of expression of an X-linked genetic condition in a female because of variation between individuals in the proportion of cells in which the mutant gene has been inactivated. Some females who

FIGURE 3.12

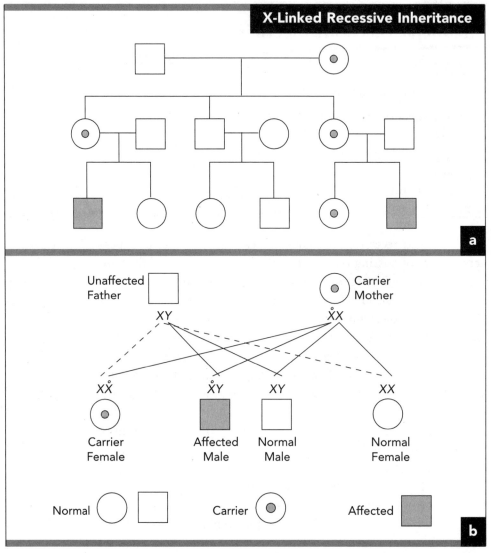

(a) An example of a family pedigree for an X-linked recessive inherited condition. (b) In X-linked recessive conditions, a carrier female has a 50% chance with each pregnancy for passing on her X chromosome with the gene mutation (X̊) and a 50% chance for passing on her normal X chromosome.

carry the gene for Duchenne muscular dystrophy, an X-linked recessive condition, for example, may exhibit signs of muscle weakness. Table 3.4 provides features of X-linked inheritance. Examples of X-linked recessive inherited conditions are listed in Table 3.5.

TABLE 3.4

Features of X-Linked Inheritance

RECESSIVE INHERITANCE	DOMINANT INHERITANCE
Vertical occurrence in pedigrees	Vertical occurrence in pedigrees
Males predominantly affected	Males and females affected
	Males more severely affected than females in some conditions

TABLE 3.5

Examples of X-Linked Recessive Inherited Conditions

Duchenne muscular dystrophy
Hemophilia
Hunter syndrome
Protan and Deutran forms of color blindness

Multifactorial Conditions

Many birth defects, as well as many common adult-onset health conditions such as cancer and heart disease, have a multifactorial cause. Multifactorial conditions are believed to result from several gene variations and environmental influences that work together to cause birth defects or disease (Figure 3.13a). These conditions cluster in families but do not demonstrate the characteristic pattern of inheritance seen with single-gene disorders (Figure 3.13b).

Neural-tube defects, for example, including spina bifida, occur as a result of genetic and environmental factors that come together during early embryonic development and cause incomplete closure of the neural tube. Research studies have demonstrated that folic acid taken prior to conception and during the first 3 months of pregnancy reduces the recurrence of neural-tube defects in women

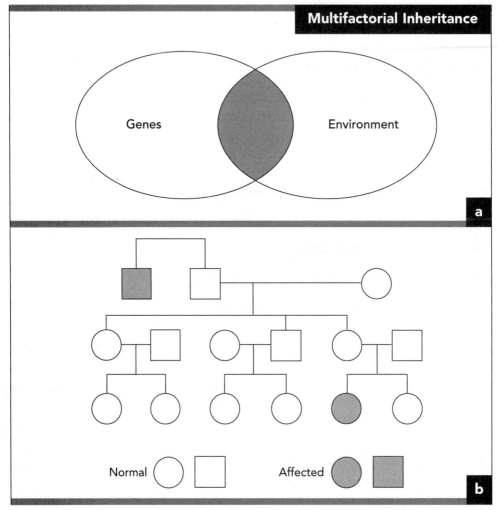

(a) Multifactorial conditions occur as a result of genetic and environmental factors combining, as represented by the area in which the two ovals overlap. (b) Family pedigree of multifactorial inheritance. Multifactorial conditions may recur in families, but they do not demonstrate the characteristic pattern of inheritance seen with single-gene conditions.

who have had a previously affected pregnancy. Folic acid is therefore an important environmental influence that appears to play a critical role in fetal development and may influence the outcome of a genetic susceptibility to neural-tube defects. Common multifactorial conditions are listed in Table 3.6.

TABLE 3.6

Common Multifactorial Conditions

Cleft lip or cleft palate
Congenital heart defects
Congenital hip dislocation
Diabetes
High blood pressure
Neural tube defects: anencephaly and spina bifida

Chromosomal Conditions

Genetic conditions involving alterations in the number or structure of chromosomes are a major category of genetic disorders. Chromosomal disorders play a causative role in birth defects, mental retardation, and malignancies. They account for more than half of all spontaneous first-trimester pregnancy losses and occur in approximately 0.7 percent (1 in 160) of live-born infants. Chromosomal abnormalities may involve one or more of the autosomes, sex chromosomes, or both. The most common type of clinically significant chromosomal alteration involves *aneuploidy,* an abnormal chromosome number caused by an extra or missing chromosome. Aneuploidy is always associated with mental or physical disability or both.

The most common cause of aneuploidy is nondisjunction, a failure of paired chromosomes to separate during meiosis. Women who are older than 35 years have been observed to be at higher risk for aneuploid pregnancies and, for this reason, they are offered prenatal diagnosis. Down syndrome is an example of aneuploidy. People who have Down syndrome have a complete extra chromosome number 21. The extra chromosome 21 causes a person with Down syndrome to have a characteristic facial appearance, an increased risk for certain health problems, such as congenital heart disease and thyroid, hearing, and vision problems, and varying degrees of mental disability (Figure 3.14a and b). Turner syndrome is another example of aneuploidy in which a single X chromosome is missing. Examples of common chromosomal conditions are listed in Table 3.7.

FIGURE 3.14

(a) *The karyotype of a person with trisomy 21 (Down syndrome) contains three number 21 chromosomes, one parent having passed on two number 21 chromosomes instead of the usual one. (Reprinted with permission from David H. Ledbetter, PhD, FACMG.) (b) A 5-year-old boy who has Down syndrome.*

TABLE 3.7

Common Chromosomal Conditions

Down syndrome (trisomy 21)
Trisomy 13
Trisomy 18
Turner syndrome (45, X0)
Klinefelter syndrome (47, XXY)

A chromosomal disorder may also occur as the result of a structural rearrangement within or between chromosomes. Chromosomal rearrangements are less common than is aneuploidy but occur in approximately 1 in 500 newborns. A chromosomal rearrangement is said to be *balanced* if the chromosome set appears to contain all of the correct chromosomal material but arranged in an unusual way. A chromosomal rearrangement is *unbalanced* when there is additional or missing chromosomal material, such as a duplication or a deletion of a chromosome segment; for example, in Figure 3.15a, a portion of chromosome 5 is missing, as shown by the arrow. In such situations, the normal balance of functional genes is disturbed, and mental and physical development is thus altered.

A person who carries a balanced chromosomal rearrangement does not usually have any associated mental or physical disabilities. A carrier, however, has an increased chance of experiencing pregnancy loss or of having children with an unbalanced chromosomal constitution that results in mental or physical disabilities. An example of a common chromosomal translocation is shown in Figure 3.15b, which shows a balanced rearrangement of chromosomes 14 and 21. In this situation, one chromosome 21 has been rearranged (translocated) to chromosome 14, as shown by the arrows. A person who carries this particular chromosome rearrangement has an increased risk for having children with Down syndrome (an extra chromosome 21), which is inherited as a result of the parent being a carrier of a balanced chromosomal rearrangement.

FIGURE 3.15

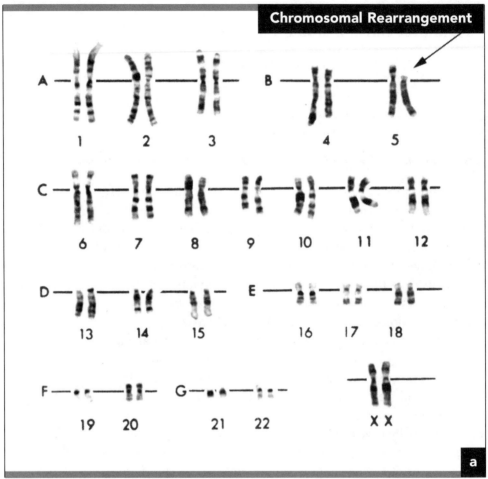

(a) An unbalanced chromosome arrangement with a portion of chromosome 5 missing.

Other Recognized Patterns of Inheritance

Although most observed patterns of inheritance for single-gene disorders follow principles of Mendelian inheritance, analysis of certain genetic conditions on both the clinical and molecular levels has demonstrated several other modes of inheritance. Nonclassical mechanisms of inheritance include mitochondrial inheritance, mosaicism, genomic imprinting, and uniparental disomy. These mechanisms may affect the transmission or expression of single-

FIGURE 3.15

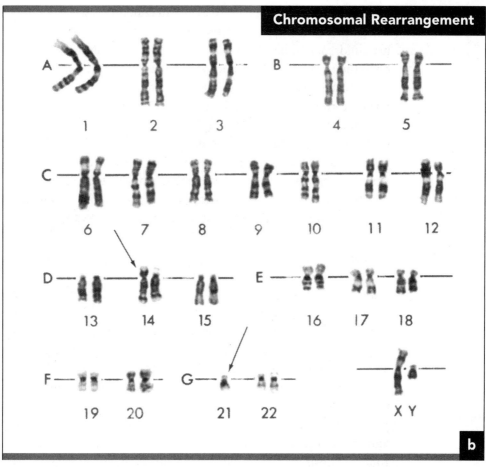

(b) Balanced translocation of chromosomes 14 and 21. (Reprinted with permission from David H. Ledbetter, PhD, FACMG.)

gene disorders or may work independently to result in physical or mental disabilities.

Mitochondrial Inheritance

Gene mutations of mitochondrial DNA (mtDNA) have been identified as the cause of a number of neuromuscular disorders that do not follow a typical Mendelian inheritance pattern. Numerous copies of mtDNA are packaged in a circular chromosome that is present *outside* the cell nucleus, in the cytoplasm; mtDNA is inherited only through the mother, because the ovum contains numerous mitochondria, whereas sperm contain very few and these do not persist in offspring. A mother passes her mtDNA on to all of her children, and her daughters, in turn, pass the same information on to their children, though her sons do not. In this pattern of inheritance, called *maternal inheritance,* affected individuals are always related through the maternal line.

Mitochondrial disorders display variable expression, probably because of the presence of both mutated mtDNA and normal mtDNA in the cells of an individual. A condition called *Leber's hereditary optic neuropathy* is one example of a mitochondrial disorder. In affected people, rapid loss of central vision in both eyes occurs in young adulthood because of optic-nerve death.

Mosaicism

Mosaicism is defined as the presence of two or more cell lines in an individual or body tissue that, although genetically different, have evolved from the same zygote (fertilized egg). Mosaicism for single-gene mutations can be either somatic (i.e., present in the cells of the body) or germline (i.e., limited to the gonads). Mosaicism provides an explanation for several unique clinical observations.

Somatic mosaicism may occur in body cells during embryonic development yet manifest itself in only one segment of the body. A person with neurofibromatosis type 1 may have unaffected parents and only a segment of that person's body may be affected with neurofibromatosis. If the mosaicism occurs at an early stage in fetal development, germ cells may also be affected. In this situation, the person would have reproductive cells that contain the gene, and the condition would be transmissible to offspring.

Germline mosaicism occurs during early development of the germ cells (egg and sperm), allowing populations of both normal and abnormal eggs or sperm to exist in a person who does not exhibit the condition. People with germline mosaicism for a condition such as autosomal-dominant osteogenesis imperfecta may be physically unaffected yet have an increased chance of giving birth to children with an autosomal-dominant form of osteogenesis imperfecta because of the presence of multiple eggs or sperm containing the mutation.

Chromosomal Mosaicism

Sometimes a person has chromosomal mosaicism, in which the body cells contain two or more different chromosomal numbers or arrangements. The most common cause of chromosomal mosaicism is a failure of chromosomes to divide properly in early embryonic development. A developing female embryo, for example, might lose an X chromosome during early cell division and continue to develop two cell lines: one line will have a complete set of 46 chromosomes and two X chromosomes, and the other will have only 45 chromosomes and a single X chromosome (Figure 3.16). The result, a condition called *mosaic Turner syndrome,* could be associated with varying degrees of short stature and infertility. The effects of chromosomal mosaicism may be difficult to assess and depend on the timing of the event of nondisjunction, the chromosomes involved, and the body tissues affected.

FIGURE 3.16

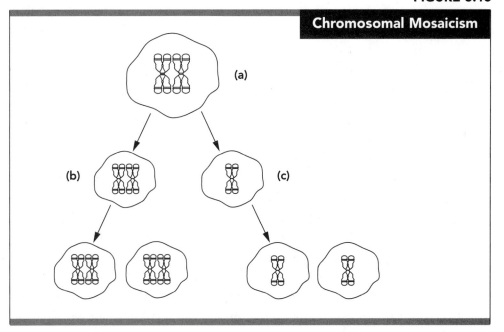

Chromosomal mosaicism is the presence of body cells that contain two or more different chromosomal numbers or arrangements. A developing female embryo (a) might lose an X chromosome during early cell division and continue to develop two cells lines. One will have a complete set of 46 chromosomes and two X chromosomes (b), and the other will have only 45 chromosomes and a single X chromosome (c).

Genomic Imprinting

In a considerable number of genetic disorders, the expression of the genetic condition depends on whether the altered gene has been inherited from the mother or the father. This phenomenon is called *imprinting*. Two examples of imprinting are early-onset myotonic dystrophy, which occurs when the gene mutation is inherited through the mother, and early-onset Huntington's disease, in which the gene mutation is inherited through the father. Prader-Willi syndrome and Angelman syndrome provide two other clinical examples of genomic imprinting. Prader-Willi syndrome, a relatively common genetic condition characterized by obesity, small hands and feet, short stature, and mental retardation is sometimes an example of uniparental disomy. Many people who have Prader-Willi syndrome have an observed deletion on chromosome number 15, inherited from the father. They therefore have genetic information on chromosome 15 that comes only from their mothers. A smaller number of people with Prader-Willi syndrome have no observable deletion. Further study of those people without the chromosomal deletion has revealed that some of these individuals have two intact number 15 chromosomes but that both members of the chromosome pair have been inherited from the mother and neither from the father. A genetic condition called *Angelman syndrome* emerges from a very different pattern. Those with this syndrome, characterized by mental retardation and distinct physical and facial features, have a deficiency on a region of chromosome 15 inherited from their mothers; they have genetic information on chromosome 15 from their fathers only. This unusual finding has led scientists to suspect that normal human development requires that one or more genes on chromosome number 15 must be inherited from each parent.

Uniparental Disomy

Modern molecular genetic techniques have allowed the parental source of chromosomes to be identified, which has helped to uncover the origin of a number of previously unexplained genetic disorders. *Uniparental disomy* has been described as the presence of a cell line containing two chromosomes of a given kind, inherited from one parent. Father-to-son transmission of hemophilia has been documented, for example. In this situation, an affected boy inherits both his X and his Y chromosomes from his father and no X chromosome from his mother.

CONCLUSION

The field of molecular genetics is expanding rapidly, and new information and insights are being gained about the genetic contribution to health and disease. Familiarity with basic genetic principles and patterns of inheritance will help nurses assist in recording, interpreting, and explaining genetic information to patients and their families. Knowledge of genetics will further enable nurses to identify individuals and families in need of genetic evaluation and counseling and to provide current genetic information.

SUMMARY POINTS

Basic genetic principles

- Genes: the units of heredity
- DNA: the hereditary material, composed of sugar, phosphate, and nitrogenous bases and shaped like a double helix
- Chromosomes: the packages of genes
- Cell division: how the body grows
  - Meiosis: formation of reproductive cells (gametes)
  - Mitosis: replication of cells to form two identical daughter cells
- DNA, RNA, Protein: the central dogma

Genes: a critical component of health and disease

- Gene mutations
- Germline mutations: inherited
- Somatic mutations: within specific body tissues

Classification of genetic disorders

- Autosomal dominant
  - Vertical transmission
  - Fifty percent chance for recurrence

- Variable expression among family members
- Reduced penetrance
- Autosomal recessive
  - Horizontal pattern of inheritance
  - More common among certain ethnic groups and in the presence of genetic relatedness
  - Twenty-five percent chance for an affected child; 50 percent chance for carrier children; 25 percent chance for child without gene
- X-linked recessive
  - Transmitted by females to males
  - Risk of recurrence for female carrier: 25 percent chance of having carrier daughters; 25 percent chance of having affected sons
  - May manifest in female carriers
  - Effects of Lyonization
- Multifactorial inheritance
  - Genes plus environment
  - Threshold effect
  - No specific pattern of inheritance but may recur in families

Chromosomal genetic conditions

- Extra or missing chromosomes
- Effects on mental or physical development
- Chromosomal mosaicism

Other recognized patterns of inheritance

- Mitochondrial inheritance: genes inherited through the mother
- Mosaicism: presence of genetically different cell lines that have evolved from the same zygote but that differ genetically because of a mutation following conception or nondisjunction
- Genomic imprinting: expression of condition dependent on whether gene has been inherited from mother or father
- Uniparental disomy: pairs of genes inherited from same parent

Applications to nursing practice

- Identifying individuals and families in need of further genetic information
- Providing appropriate genetic information
- Making referrals to appropriate genetic resources
- Providing timely and knowledgeable support to families identified as being at risk for genetic conditions

QUESTIONS FOR CRITICAL THINKING

1. You are asked to give a talk about genetics to a fifth-grade class. How would you describe the structure and function of the following: a gene; DNA; a chromosome?

2. In recording the family history of a 40-year-old male patient, you note that the patient's father died at age 50 from dementia; his sister, age 48, has the early symptoms of dementia; a paternal aunt, age 66, is in a nursing home and has been diagnosed with dementia; a paternal uncle is known to have died in his fifties from dementia; and the patient has three sons, all of whom are in their teens. Draw a family pedigree. What information might you give to the patient?

3. State three expected outcomes of new genetic technologies designed to map and sequence human genes. Describe whether you believe these outcomes will be helpful in the diagnosis and treatment of human disease. How would you explain to one of your patients the genetic contribution to health and illness?

REFERENCES

Bennett, R., et al. (1995). Recommendations for standardized human pedigree nomenclature. *J. Genet. Counsel.* 4(4):267–279.

Harrigan, J. (1996). Nursing practice management: muscular dystrophy. *J. School Nurs.* 12(2):38–40.

Iron Overload Diseases Association, Inc. (IOD), 433 Westwind Drive, North Palm Beach, FL 33408. Telephone: 407-840-8512; fax: 407-842- 9881.

Keller, J.F., and Blausey, L.A. (1992). Diagnosis, treatment, and nursing management of the patient with hemochromatosis. *Oncol. Nurs. Forum* 19(9):1359–1365.

Levitan, M. (1988). *Textbook of human genetics* (3rd ed.). New York: Oxford University Press.

Lewin, B. (1994). *Genes V.* Oxford: Oxford University Press.

Lyon, M.F. (1962). Sex chromatin and gene action in the mammalian X chromosome. *Am. J. Hum. Genet.* 14:135–148.

McKusick, V.A. (1994). *Mendelian inheritance in man: a catalogue of human genes and genetic disorders* (11th ed.). Baltimore: Johns Hopkins University Press.

Muscular Dystrophy Association (MDA), 3300 East Sunrise Drive, Tucson, AZ 85718-3208. Telephone: 602-529-2000; fax: 602-529-5300.

National Down Syndrome Congress (NDSC), 1605 Chantilly Drive, Suite 250, Atlanta, GA 30324-3269. Telephone: 1-800-232-6372 or 404-633-1555; fax: 404-633-2817.

National Institutes of Health, National Cancer Institute (1995). *Understanding gene testing* (NIH pub. no. 96-3905). US Department of Health and Human Services.

Rimoin, D., Connor J.M., and Pyeritz, R. (eds.) (1996). *Emery–Rimoin principles and practice of medical genetics* (3rd ed.). New York: Churchill Livingstone.

Stray-Gundersen, K. (1986). *Babies with Down syndrome.* Washington, DC: Woodbine House.

Thompson, M.W., McInnes, R.R., and Willard, H.F. (1991). Thompson & Thompson: *Genetics in medicine* (5th ed.). Philadelphia: Saunders.

Watson, J.D., and Crick, F.N.C. (1953). Molecular structure of nucleic acids: a structure for deoxyribonucleic acid. *Nature* 171:737–738.

Williams, J.K. (1996). Genetic issues for perinatal nurses. March of Dimes Nursing Modules. White Plains, NY: March of Dimes Birth Defects Foundation.

INTERPRETING PATTERNS OF INHERITANCE: DUCHENNE MUSCULAR DYSTROPHY

Ms. M is an 18-year-old, single patient who comes to the family health clinic for her annual physical examination. During the review of family history with the nurse, Ms. M explains that Duchenne muscular dystrophy (DMD),* a severe, degenerative muscle disorder, was recently diagnosed in her younger brother, age 4. The interview revealed that she had a maternal uncle who died at age 16 from complications associated with DMD. Ms. M has two older sisters, ages 20 and 21, and one other brother, age 12. She and her sisters have a different father than her two younger brothers. Ms. M states that her family, especially her mother, is having difficulty with her brother's diagnosis; she explains, "My mother took care of my uncle at the time of his death." Ms. M asks the nurse about her chances for having children with DMD.

*DMD is a severe, degenerative muscle disorder that affects primarily males. Affected boys develop normally during the first year or two of life but develop muscle weakness at 3–5 years, when they may begin to have difficulty climbing stairs and rising from a sitting position. People who have DMD usually need a wheelchair by age 12 and do not live beyond age 20. DNA testing is currently available for diagnosis, carrier status, and prenatal diagnosis (Harrigan, 1996; Rimoin et al., 1996; Muscular Dystrophy Association).

- What can the nurse tell Ms. M about the pattern of inheritance of DMD in her family?
- What information about DMD might the nurse discuss with Ms. M?
- How can the nurse explain the genetic implications of DMD to Ms. M?

Discussion

The nurse's role in relation to Ms. M (and other such patients) includes educator, counselor, and referral source. All explanations about the disorder, its pattern of inheritance, and implications for the patient and her family should be given in words suitable for the patient's level of education and emotional availability. Because patient education is a process, information may have to be given several times—and possibly on several different occasions and using different methods—before the patient has a full understanding of the issues involved.

Before providing information about the condition, the nurse should determine the patient's current understanding of DMD and its course. The nurse should also describe DMD as a severe and degenerative condition and should discuss the symptoms and disease progression, letting the patient know what is possible and what is probable. Patients are often then most interested in patterns of inheritance and how those patterns affect them and their families. In this case, the nurse can let

Ms. M know that because of her family history of DMD, she has an increased chance of being a carrier.

The nurse might provide information about inheritance of DMD in Ms. M's family in the following way: DMD is often inherited in families; approximately one-third of persons who develop DMD do so as a result of having been born with a new gene mutation; approximately two-thirds of persons with DMD have mothers who carry the gene on one of their two X chromosomes. This is called *X-linked inheritance*. Review of the family history reveals that DMD probably is inherited in Ms. M's family, because Ms. M's mother has a brother who had DMD and a son who also has it. Therefore, Ms. M's mother is a carrier, and Ms. M has a 50 percent chance of being a carrier. If Ms. M is a carrier, then she also has a 50 percent chance of having a son who has DMD and a 50 percent chance of having a daughter who is a carrier.

Because the gene for DMD has been identified and characterized, carrier testing and prenatal diagnosis are now available using DNA testing and can be offered to patients and their families. The nurse can explain the availability and helpfulness of genetic counseling for further evaluation and discussion of Ms. M's family history and concerns. In collaboration with the physician, the nurse can refer Ms. M to genetic specialists for more information about DMD, her chances of being a DMD carrier, and available carrier and prenatal diagnostic testing. The nurse could also suggest the brother's physician as an information source and the Muscular Dystrophy Assocation as a source of both information and support.

Another nursing function might be to explore family dynamics with the patient to identify family concerns. In this setting, Ms. M provides some information about her mother's response to the diagnosis of DMD in her brother. The nurse is aware that diagnosis of a genetic condition in a family has an impact on all family members, who may be at different stages of understanding, acceptance, and integration of new genetic information into their self-concept and views of their family. If the patient is open to discussion and expresses concerns, the nurse can offer other local resources, including psychosocial support services for the patient and her family. A referral to other families who have similarly affected family members is often helpful and could be arranged through the Muscular Dystrophy Association or in collaboration with genetics professionals at the genetics clinic.

INTERPRETING PATTERNS OF INHERITANCE: HEREDITARY HEMOCHROMATOSIS

Mr. C is a 45-year-old patient at the ambulatory care clinic. He is currently undergoing weekly phlebotomy (blood drawing) as treatment for hereditary hemochromatosis (HH).* During his treatment, he tells the nurse that he has two children and that, although it had been suggested that the children be tested, the parents have not followed through with the suggestion. He says that he is concerned about the children's chances of developing HH. When questioned about other family history, he reveals that one of his first cousins on his mother's side of the family also has HH and is being treated in a clinic in Florida. The nurse offers to locate for Mr. C information about HH that might help his family.

*HH is a common single-gene disorder occurring among Caucasians in 1 in every 400 people, with a carrier frequency of 1 in 10 people. Symptoms of HH usually develop in mid-adulthood. Excessive iron accumulates in various body organs, including the pancreas, liver, and heart, causing diabetes, cirrhosis of the liver, liver cancer, and cardiac dysfunction, and often leading to early death. Screening and diagnostic testing are available for HH. Treatment is simple and effective and involves regular phlebotomy to deplete the excess surplus of iron. If HH is diagnosed early enough and is treated appropriately, life expectancy is normal (Keller & Blausey, 1992; Rimoin et al., 1996; Iron Overload Diseases Association, Inc.).

- If you were the nurse, how would you describe the pattern of inheritance for HH in Mr. C's family and explain how the condition occurred in his cousin but not in other family members?
- What information could you relate about the chance of Mr. C's children developing HH?
- What additional resources can you offer to Mr. C and his family to help answer his questions?

Discussion

The nurse should first explore and assess Mr. C's current understanding of HH and its inheritance in families. In her roles as educator and referral source, the nurse could offer information about HH in the following way: HH is inherited in families when two parents are carriers of a single altered gene. A person who is a carrier does not usually experience any of the symptoms of HH but when two carrier parents have children, they have a 25 percent chance of each passing on the gene for HH to a child, who would inherit two recessive genes (a "double dose" of the gene). A person who inherits two genes for HH will develop the symptoms of the condition and should be offered treatment. Testing to determine whether a person has inherited two altered genes should be offered to all those with a family history of HH, to help identify those who need to be treated. Because of Mr. C's personal and family history of HH, his children are more likely than others to be affected.

In inherited conditions such as HH, one or more family members might have the condition in the same generation. This is because the gene for HH may be passed down through generations as a *trait*, which would not cause HH, because a carrier of a gene for HH also carries a normal version of the gene.

While performing Mr. C's phlebotomy, the nurse has the opportunity to explore Mr. C's relationship with his children and to identify appropriate ways to help Mr. C

obtain information about testing and treatment of HH for his children. The nurse also could offer to provide Mr. C and his family with information about national support organizations for HH and might relate that support for families who have genetic conditions such as HH is now widely available and is often very helpful. She can also suggest that Mr. C speak with his physician about a referral to a genetics clinic, which might provide a forum for further evaluation of family history and discussion of HH and the implications for other family members.

THE EFFECT OF CHROMOSOMAL ABNORMALITIES: DOWN SYNDROME

The nurse attends the birth of Ms. N's baby. The 19-year-old Ms. N presented late for prenatal care and did not undergo prenatal genetic counseling or testing during her pregnancy. The nurse notes that, at birth, Ms. N's baby has low muscle tone and low-set ears and that the corners of the baby's eyes are upslanted. A heart defect is also noted. The nurse and the attending physician suspect that the baby has Down syndrome. The nurse and physician explain to Ms. N that her baby has features that are suggestive of a condition called *Down syndrome*.* To confirm this

diagnosis, a special blood test known as a *chromosome analysis* is needed. With Ms. N's permission, the nurse obtains a blood sample from the baby, and this sample is sent to a local laboratory for analysis. Ms. N asks the nurse to explain what the test will tell her, and what Down syndrome is. Ms. N says that no one else in her family has had a baby with Down syndrome and that she thought that it happened only if the mother was "older."

*Down syndrome is a genetic condition that occurs in approximately 1 in 800 live births. It is one of the most common causes of mental retardation and disability. People who have Down syndrome have varying degrees of mental retardation. Approximately 40 percent of people who have Down syndrome have a heart defect. Other

- If you were the nurse, what additional information would you provide about the chromosome analysis?
- What other information might you discuss about Down syndrome?
- What can be said about the inheritance of Down syndrome in most families?

Discussion

In this setting, the nurse recognizes that the possibility of Down syndrome in Ms. N's baby may cause anxiety and concern and considers the possibility of difficulty with coping as a potential response to this information. Her main activities include supportive care and provision of accurate information in language and a format that Ms. N will be able to comprehend.

The nurse explains that the chromosome analysis is a blood test that can pro-

associated medical conditions include hypothyroidism, visual problems, and an increased risk for leukemia and Alzheimer's disease.

The diagnosis of Down syndrome is made by chromosomal analysis, which reveals three number 21 chromosomes present in all analyzed cells. For chromosomal analysis, a blood sample from the baby must be obtained in a tube containing sodium heparin to prevent the blood's clotting. White blood cells are collected from the sample, placed in tissue culture, and stimulated to grow. Cells are stained and treated with a solution to release the chromosomes from the nucleus. Chromosomes are then fixed and spread on slides and stained for analysis. Results may be available in 5–7 days, but a rapid test can provide preliminary diagnostic information within 2–3 days (Stray-Gundersen, 1986; Thompson et al., 1991; National Down Syndrome Congress).

vide information about Ms. N's baby's chromosomes, the packages of genes inside all cells of the body. She relates that there are usually 46 chromosomes in body cells, but sometimes a person is born with too many or too few. People who have Down syndrome have three number 21 chromosomes. The presence of the extra chromosome 21 can cause low muscle tone and heart defects. When the test results are back, Ms. N's pediatrician and a professional who specializes in genetics will be available to explain the results and to answer her questions.

To answer Ms. N's question regarding the inheritance of Down syndrome in families, the nurse might explain that Down syndrome occurs in all populations. Any woman who becomes pregnant has a chance of having a baby with Down syndrome, but for women who are 35 or older during pregnancy, this chance increases. In most families, Down syndrome is not inherited.

Ms. N could be supported further as the nurse explores her other questions and concerns and explains that if she (the nurse) is not able to answer Ms. N's questions, she can help Ms. N to locate a professional who *can* provide answers. The nurse might discuss with Ms. N that if the test results come back positive (i.e., showing that her baby does have Down syndrome), many resources are available to help Ms. N learn more about this condition. Among these resources are genetics professionals, support groups, other parents who have children with Down syndrome, and written materials available through national Down-syndrome organizations. Other resources to support the medical, social, and emotional needs of parents of children who have Down syndrome are also available.

Integrating Genetics into Nursing Practice

Neither [medicine nor surgery] can do anything but remove obstructions; neither can cure; nature alone cures. Surgery removes the bullet out of the limb, which is an obstruction to cure, but nature heals the wound . . . And what nursing has to do in either case, is to put the patient in the best possible condition for nature to act upon him. Florence Nightingale

OBJECTIVE

Apply knowledge of genetics to nursing practice.

RATIONALE

As more is learned about the contribution of genetics to health and disease, nursing roles should expand in the delivery of genetic health care in all clinical settings.

APPLICATION ACTIVITIES

- Help with collecting, reporting, and recording genetic information.
- Offer genetic information and resources to patients and families.
- Support informed choice regarding health and reproductive decisions.
- Advocate for privacy, confidentiality, and nondiscrimination with regard to genetic information.
- Participate in management of patients with genetic conditions.

GENE CARE LINK

For more information about human genes and genetic disorders, go to http://www. jbpub.com/ clinical-genetics.

INTRODUCTION

Nurses have long provided genetics-related nursing care, in a variety of settings in the health care system and the community, to patients and families with genetic conditions. One ongoing nursing responsibility is recognizing patients who could benefit from genetic services. Those services might include providing genetic information, counseling, testing, or referral for genetic counseling and evaluation. Another ongoing responsibility is awareness of sources of genetic information especially suitable for answering patients' questions regarding genetic implications. This chapter focuses on additional *nursing practice activities that are evolving as a result of information derived from the Human Genome Project.*

Nursing practice in genetics incorporates the principles of human genetics into patient care. The practice is carried out in collaboration with other health professionals, including genetic specialists, to assist with health improvement, maintenance, and restoration (Fibison, 1983; Forsman, 1994). Genetics-related nursing practice includes care of patients who have genetic conditions, individuals who may be predisposed to develop or to pass on a genetic condition, or people who are seeking genetic information and referral for additional genetic services.

In any setting, five main nursing activities have evolved in genetics nursing practice: (1) helping with collecting, reporting, and recording genetic information; (2) offering genetic information and resources to patients and families; (3) supporting informed choice regarding health and reproductive decisions; (4) advocating for privacy, confidentiality, and nondiscrimination with regard to genetic information; and (5) participating in management of patients with genetic conditions (Scanlon & Fibison, 1995).

COLLECTING, REPORTING, AND RECORDING GENETIC INFORMATION

Genetic information is a unique body of information about the inheritable contributions to a person's health. It includes identification of a specific genetic trait or inherited condition, DNA diagnosis of a genetic condition, and determination of the presence of a clinical genetic condition. Family history, physical examination, and genetic screening and testing all are sources of genetic information (Scanlon & Fibison, 1995).

To identify and provide genetic information that may be helpful to patients and families, nurses can expand their scope of practice to include assessing genetic health, offering genetic information that takes into account ethnocultural and educational differences, and offering and describing genetic screening and testing. Nurses can also support patients and families with genetic concerns by addressing issues surrounding the management of genetic information, such as assuring informed health choices and advocating for privacy and confidentiality of genetic information and for nondiscrimination.

Genetic health assessment is a continuous process of collecting information by which to identify individuals and families with actual or potential genetic health concerns who may benefit from further genetic information, counseling, and evaluation. This process can begin prenatally and continue through the life cycle and even after death. Nurses can conduct various aspects of the assessment to obtain genetic health information; they can evaluate family and past reproductive history, including maternal health history, identify ethnic background, and conduct a physical assessment. Genetic health assessment always involves determining a patient's or family's understanding of actual or potential genetic health concerns and identifying the nature of communication regarding these issues within a family (Williams and Lea, 1995; Lea, 1995; Williams, 1996).

Family and Reproductive History

Nurses can obtain family and reproductive histories at an initial primary care visit and regularly update them to assess for additional genetic health issues. The nurse's goal is to gather pertinent information about a patient and family that might provide clues that a genetic trait or inherited condition is present. A genetic-health-history questionnaire can be used to identify specific genetic conditions for which further information, evaluation, and testing may be offered (Figures 4.1 and 4.2). The nurse can then determine whether further genetic evaluation or testing should be offered for the trait or condition in question. When an inherited condition is suspected, the nurse constructs a family tree or pedigree to determine the implications for the patient and other family members. A detailed and accurate family history, as outlined in Chapter 3, provides the most complete family genetic health information. The family history should cover at least three generations (grandparents, parents, full and half-siblings, aunts, uncles, and cousins) and should include information about current and past health status of all family members, including age of onset of any illness and cause of and age at death.

When ascertaining the health status of family members, the nurse should ask about medical conditions known to have a heritable component and for which genetic testing may be offered (Table 4.1). The nurse could also obtain as part of the genetic family assessment information regarding the presence of birth defects, mental retardation, familial traits, or similarly affected family members.

The nurse records the presence of genetic relatedness (consanguinity). All persons are believed to have at least six to eight gene mutations that, when present in both parents, may lead to a recessive condition in children. Individuals with common ancestors share more genes than do those who are unrelated. The number of shared genes depends on the degree of relationship. A father and daughter, for example, share one-half of their genes, whereas first cousins share one in eight genes. Knowledge of genetic relatedness allows the nurse an opportunity to offer additional genetic counseling and evaluation. It may also provide an explanation for families that experience the birth of a child with a rare autosomal-recessive condition (Harper, 1993).

The nurse should ask specific questions regarding reproductive history: history of miscarriage or stillbirth to assess for possible chromosomal conditions such as an inherited translocation; history

FIGURE 4.1

ProgramME Genetic History Questionnaire for Prenatal Patients

- The answers to these questions will help in the care of your pregnancy.
- Please answer these questions as well as you can. All answers will remain private.
- If you need help answering the questions, please ask.

1. When your baby is born, will you be 35 years of age or older?   ☐ No   ☐ Yes

Where your ancestors came from may sometimes give us important information about the health of your baby.

2. Is your family . . .
   . . . from Southeast Asia, Taiwan, China, or the Philippines?   ☐ No   ☐ Yes   ☐ Not sure
   . . . from Italy, Greece, or the Middle East?   ☐ No   ☐ Yes   ☐ Not sure
   . . . African-American (Black)?   ☐ No   ☐ Yes   ☐ Not sure
   . . . Hispanic/Puerto Rican?   ☐ No   ☐ Yes   ☐ Not sure
3. Is your family, or your baby's father's family, Eastern European (Ashkenazi) Jewish?
   ☐ No   ☐ Yes   ☐ Not sure

The next nine questions will be about you, your baby's father, and both of your families. When we say "blood relative," we mean your child (or unborn baby), mother, father, sister, brother, grandparent, aunt, uncle, niece, nephew, or cousin.

4. Were you, or your baby's father, or any blood relative born with an opening in the back or spine, also called spina bifida?   ☐ No   ☐ Yes   ☐ Not sure
5. Was there ever a baby (or unborn baby) in your family or your baby's father's family who had an opening in the head, also called anencephaly?   ☐ No   ☐ Yes   ☐ Not sure
6. Is any blood relative in your family or your baby's father's family mentally retarded?
   ☐ No   ☐ Yes   ☐ Not sure
7. Have you, or your baby's father, or any blood relative had an unborn baby or a child who had Down syndrome (some call it trisomy 21)?   ☐ No   ☐ Yes   ☐ Not sure
8. Do you, or your baby's father, or any blood relative have any other chromosome problem?   ☐ No   ☐ Yes   ☐ Not sure

Patient's Name: _____   **Date of Birth:** _____/_____/_____

(cont.)

FIGURE 4.1 (cont.)

ProgramME Genetic History Questionnaire for Prenatal Patients (cont.)

9. Do you, or your baby's father, or any blood relative have:
   a. ... cystic fibrosis (CF)?   ☐ No   ☐ Yes   ☐ Not sure
   b. ... fragile X syndrome?   ☐ No   ☐ Yes   ☐ Not sure
   c. ... muscular dystrophy?   ☐ No   ☐ Yes   ☐ Not sure
   d. ... hemophilia or other bleeding disorder?   ☐ No   ☐ Yes   ☐ Not sure
   e. ... Huntington's disease?   ☐ No   ☐ Yes   ☐ Not sure
10. Were you, or your baby's father, or any blood relative born with:
   a. ... a heart defect?   ☐ No   ☐ Yes   ☐ Not sure
   b. ... a cleft lip or palate?   ☐ No   ☐ Yes   ☐ Not sure
   c. ... any other birth defect?   ☐ No   ☐ Yes   ☐ Not sure
11. Have you ever had:
   ... two or more miscarriages?   ☐ No   ☐ Yes
   ... a stillborn baby *and* one or more miscarriage(s)?   ☐ No   ☐ Yes
12. Do you, or your baby's father, or any blood relative have any other disease or health problem that is inherited (passed on in the family)?   ☐ No   ☐ Yes   ☐ Not sure

The next three questions will be about medical conditions that you (the patient) may have.

13. Do you have diabetes?   ☐ No   ☐ Yes
14. Do you have, or have you ever had treatment for, PKU (phenylketonuria) or hyperphenylalaninemia (hyperphe)?   ☐ No   ☐ Yes   ☐ Not sure
15. During this pregnancy, have you taken:
   a. ... medications for seizures?   ☐ No   ☐ Yes
      (examples are Dilantin, valproic acid, Depakene, Tegretol, Atretol, Mysoline, Tridione)
   b. ... lithium for depression?   ☐ No   ☐ Yes
      (examples are Eskalith, Lithobid, Lithonate)
   c. ... pills (Accutane, isotretinoin) for acne?   ☐ No   ☐ Yes

Completed by: _____   **Date:** ____/____/_____

Source: Foundation for Blood Research and Project No. MCJ-231003-02 from the Maternal and Child Health Bureau ([Title V, Social Security Act], Health Resources and Services Administration, Department of Health and Human Services.)

FIGURE 4.2

Genetic Family History Questionnaire for Patients and Families

Do you or does any family member in either of your families have (include any relatives):

	YES	NO
Cleft lip or cleft palate?	_____	_____
Heart problem (from birth)?	_____	_____
Spina bifida (opening in the spine)?	_____	_____
Abdominal wall opening (opening in belly at birth)?	_____	_____
Kidney problems (other than infections)?	_____	_____
Hearing loss, deafness?	_____	_____
Visual loss, blindness (not age related)?	_____	_____
Mental retardation?	_____	_____
Fragile X syndrome?	_____	_____
Muscular dystrophy (muscle weakness)?	_____	_____
Neurofibromatosis (many skin or nerve tumors on body)?	_____	_____
Cystic fibrosis (CF)?	_____	_____
Hemophilia (bleeding disorder)?	_____	_____
Hemochromatosis (iron overload; iron sickness)?	_____	_____
Sickle cell anemia?	_____	_____
Thalassemia?	_____	_____
Down syndrome (or any other chromosome problem)?	_____	_____
A history of two or more miscarriages (for yourself)?	_____	_____
A history of stillbirths or any babies who died before 1 year of age?	_____	_____
Any relatives that have been born with any type of birth defect?	_____	_____
Any relatives that have inherited or genetic problems?	_____	_____
Are there any other medical problems that are of concern to you?	_____	_____
Are you or your partner of Puerto Rican, Black, Southeast Asian, Mediterranean (i.e., Italian, Greek) or Ashkenazi Jewish ancestry?	_____	_____

If you checked yes to any of these questions, please contact a counselor.

Counselor's signature: _____ **Date:** _____/_____/_____

Patient name _____**Date of birth:** _____/_____/_____

Reprinted with permission from the Southern Maine Regional Genetics Services Program, Foundation for Blood Research.

TABLE 4.1

Examples of Genetic Conditions for Which DNA Carrier, Diagnostic, and Prenatal Diagnostic Testing Is Available

Achondroplasia	Myotonic dystrophy
Cystic fibrosis	Neurofibromatosis
Duchenne muscular dystrophy	Polycystic kidney disease
Fragile X syndrome	Sickle cell disease
Huntington's disease	Tay-Sachs disease
Hemophilia	Thalassemia

of family members with mental or physical abnormalities that may be inherited; such maternal health conditions as insulin-dependent diabetes, seizures, or maternal phenylketonuria (PKU), which may increase the chance for birth defects in children; and such exposures during pregnancy as alcohol or medications that might harm the fetus or increase the chance for birth defects (University of Colorado Health Sciences Center, 1988).

The nurse also notes maternal age. Women 35 years or older who are considering reproduction or who are already pregnant should be offered prenatal diagnosis (e.g., amniocentesis) because of the association between advancing maternal age and increased incidence of chromosomal abnormalities such as Down syndrome. Less information is available regarding the effects of paternal age and the risk for chromosomal abnormalities. Individuals with specific concerns about maternal or paternal age or who are considering prenatal diagnosis may be offered genetic counseling for further discussion of their specific reproductive age–related risks and available prenatal testing.

When assessment of family history reveals that an individual has been adopted, genetic health assessment becomes more difficult. The nurse and health care team should make all efforts to obtain as much information as possible about the biological parents, including their ethnic background.

Racial and Ethnic Background

Some genetic conditions are more common among certain racial and ethnic populations, and for a growing number of conditions, carrier testing and prenatal diagnosis are available. Recommendations by professional organizations such as the American College of Obstetrics and Gynecology (ACOG) currently stipulate that relevant racial and ethnic populations be offered carrier testing (ACOG, 1987; ACOG, 1995; ACOG, 1996a, b). Individuals of Ashkenazi Jewish ancestry, for example, are offered carrier testing to identify carriers of Tay-Sachs disease; those of African descent are offered testing to identify carriers of sickle-cell anemia; and people of Greek, Italian, or Southeast Asian descent can be tested to identify carriers of thalassemia (Table 4.2). Carrier testing is ideally offered before conception to allow carriers time to make reproductive decisions. Prenatal diagnosis is offered and discussed when both partners of a couple are found to be carriers.

TABLE 4.2

Autosomal-Recessive Inherited Conditions for Which Carrier Testing Is Offered

CONDITION	ETHNIC GROUP	CARRIER TEST
Tay-Sachs disease	Ashkenazi Jewish	Hexosaminidase A activity testing
Sickle cell anemia	African-American, Puerto Rican, Mediterranean, Middle Eastern	Hemoglobin electrophoresis
Alpha-thalassemia	Southeast Asian, African-American	Complete blood cell count with indices; hemoglobin electrophoresis
Beta-thalassemia	Italian, Greek, African-American, Southeast Asian	Complete blood cell count with indices; hemoglobin electrophoresis

Source: Abstracted from the American College of Obstetrics and Gynecology, 1987, 1995, & 1996.

Social, Cultural, and Spiritual Assessment

Consideration of a patient's and family's social, cultural, and spiritual orientations must be part of the nurse's assessment when offering, collecting, and discussing genetic information. Patients' views on the significance of genetic disorders, their role in health and illness, reproduction and abortion, and concepts of labeling and disability must also be considered. Different backgrounds determine varied interpretations and values concerning information from genetic evaluation and testing and thus influence perceptions of health, illness, and risks (Wenger, 1991). Educational background, family structure, and family decision making function in the same way. Assessment of these factors and of beliefs, values, and expectations regarding genetic information helps the nurse to provide sensitive and appropriate material about the specific genetic topic. Some cultures, for example, encourage a belief that the cause of illness is supernatural and that health means the absence of symptoms. Offering presymptomatic or carrier testing to persons who hold these beliefs may be initially unacceptable (Hoang & Erickson, 1985).

Nurses can best provide culturally sensitive and appropriate genetic information and health care when they collaborate with family, cultural, and religious community leaders, genetics services, and patient-advocacy groups. Nursing efforts to include these resources help to improve the communication of health information in a way that lessens social, ethnic, and economic barriers.

Physical Assessment

Some conditions with a genetic etiology can be identified by physical assessment. The family history may suggest a specific area for physical assessment. A family history of neurofibromatosis, an autosomal-dominant inherited condition involving tumors of the central nervous system, for example, would prompt the nurse to perform a careful assessment of closely related family members. Skin findings such as café-au-lait spots, axillary freckling, or tumors of the skin (neurofibromas) would warrant referral for further evaluation, including genetic evaluation and counseling.

TABLE 4.3

Examples of Major Congenital Malformations	
Spina bifida or anencephaly	Open abdominal-wall defect
Cleft lip or palate	Abnormal shortening of limbs
Congenital heart disease	

Certain clinical features in infants suggest the possibility of a genetic condition and the need for further evaluation. Approximately 40 percent of normal newborns have one minor physical anomaly such as extra digits, fused second and third toes, or a single palmar crease. When three or more minor anomalies are observed during assessment, the nurse, in collaboration with the health care team, should search for major malformations such as a congenital heart defect and consider whether further evaluation (such as a chromosome study) and evaluation by a genetic specialist are warranted. Examples of major congenital anomalies are listed in Table 4.3.

When a clinical feature suggestive of a genetic condition is identified, the age of the person being examined must be considered. Developmental or intellectual deterioration in an infant, for example, could suggest an inherited metabolic condition such as Hurler syndrome, a degenerative, autosomal-recessive condition caused by the absence of a specific lysosomal enzyme. In an adult, loss of physical or mental functioning might indicate Huntington's or Alzheimer's disease. Age of examination is also important in Klinefelter syndrome, a chromosomal condition in which a male has an extra X chromosome (47,XXY). At birth, a male who has Klinefelter syndrome usually appears normal. During adolescence, however, excessive breast tissue may be observed, and further evaluation, such as a chromosome analysis, should be considered (Cohen, 1984).

When a genetic condition is suspected on physical examination or as a result of a developmental, family, or reproductive assessment, the nurse, in collaboration with the health care team, may initiate further evaluation. Genetic testing and referral to a geneticist may be recommended. Indications for such recommendations are listed in Table 4.4

OFFERING GENETIC INFORMATION AND TESTING

The many different laboratory assays that provide information leading to the diagnosis of Mendelian disorders or other conditions with a known genetic contribution are called *genetic tests.* Genetic tests can help to confirm or make a diagnosis of a specific genetic condition when family history or physical examination raises suspicions. Genetic tests can also determine whether a specific genetic trait, condition, or predisposition is present in apparently healthy individuals or their offspring. Nursing participation in genetic testing will probably broaden, especially in the areas of patient education, assurance of informed health choices and consent, advocacy for confidentiality and privacy with regard to test results, and assistance for patients in understanding the complex issues involved in genetic testing (Holtzman, 1994; Lea, 1995).

Genetic Screening

Genetic screening is the application of genetic testing to specific populations, independent of a family history of a disorder (Andrews et al., 1994; Williams, 1996). Genetic screening differs from genetic diagnosis in that it merely identifies patients who are more likely to have a genetic condition. When a screening test reveals an increased chance for the condition in question, more specific diagnostic testing is offered. Genetic screening is often considered in terms of prevention and opportunities for intervention; for example, identifying people with a genetic condition, prior to the onset of symptoms, can allow for more effective treatment planning. Babies who, in the newborn period, are identified as having PKU are an important example. Another example is identifying carriers of a specific genetic trait such as thalassemia, which allows for provision of genetic counseling and reproductive choices.

TABLE 4.4

Indications for Recommending Further Genetic Evaluation and Counseling

REPRODUCTIVE HISTORY

Recurrent spontaneous pregnancy loss

Stillbirths or infant deaths resulting from unknown or genetic causes

Consideration of marriage to a blood relative

Women age 35 years and older who are considering pregnancy

Infertility

FAMILY HISTORY

Presence in a family of a condition with a known or suspected genetic component, including chromosomal, single-gene, or multifactorial condition

Physical or mental disability affecting more than one family member

Single or multiple congenital abnormalities

Familial occurrence of cancer or the occurrence of adult-type tumors in a child

Early onset of common disorders such as heart disease, breast cancer

Unexpected reactions to drugs or anesthesia

DEVELOPMENTAL HISTORY

Abnormal or delayed development

Mental retardation

Failure to thrive in infancy or childhood

Unusually short or tall stature

Abnormalities or delays in growth or body proportions

Delayed or abnormal development of secondary sex characteristics or organs

Sources: Abstracted from ACOG, 1987; Cohen, 1984; Milunsky, 1992.

Most commonly, genetic screening occurs in prenatal and new-born programs, which involve nurses in various capacities and settings. Community nurses, for example, might participate in the follow-up of patients identified to be carriers of recessive traits for sickle-cell anemia. Neonatal nurses working with newborns and infants take part in newborn screening programs for conditions such as PKU, galactosemia, and homocystinuria. Nurses caring for pregnant women are frequently involved with offering, explaining, and reporting prenatal screening tests for neural-tube defects and Down syndrome (Wright et al., 1992; Williams, 1986). In the future, genetic screening might be applied to the detection of people who are predisposed to develop such conditions as breast and colon cancer. Nurses in a variety of settings may participate in discussions of relevant information that will allow for informed decisions and opportunities for prevention of, and early treatments for, inheritable conditions.

NURSING CONSIDERATIONS IN GENETIC SCREENING AND TESTING

Nurses involved in a range of aspects of genetic screening and testing can provide sensitive and comprehensive support to patients and families if the nurses are aware of key aspects of the process. Of prime importance are the validity of the genetic test being offered, assurance of informed choice and consent, assurance of privacy and confidentiality of test results, and advocacy for nondiscrimination.

Understanding the validity of genetic tests and of testing parameters such as a test's sensitivity (the percentage of people who have the condition or trait that the test will detect) and specificity (the percentage of people who do not have the condition or trait who will have normal results) will be helpful for nurses as newer genetic tests are developed for predisposition testing. This information is generally available from laboratories that offer testing, in the medical literature, and from genetic specialists. Knowledge of these resources will help nurses to understand what genetic tests can and cannot predict. In turn, nurses can ensure that patients receive the most accurate and complete information to assist them in their deciding whether to undergo genetic testing. Genetic testing for cancer and associated issues are discussed in more detail in Chapter 9.

Assurance of Informed Consent, Privacy, and Confidentiality: Support and Advocacy

Genetic Testing

The basis of the informed-consent process is "mutual participation, respect, and shared decision-making" (Scanlon & Fibison, 1995). The focus should be on communication between the patient and the health care provider, so that what the patient wants actually occurs. The patient should come to understand, through discussion, that there is a right to accept or refuse genetic testing. The fundamental elements of the process of informed consent are outlined in Table 4.5.

To make an informed health decision, a person must be able to understand information, to consider options, to evaluate the potential risks and benefits of a genetic test, and to communicate health choices. The capacity to make an informed health decision may be inhibited by hearing or language deficits, intellectual disabilities, or

TABLE 4.5

Topics to Be Discussed During the Informed-Consent Process

Purpose of the genetic test	Available treatment and intervention options
Reason for offering testing	
Type and nature of genetic condition being tested for	Further decision making that may be needed upon receipt of test results
Accuracy of genetic test	Consent to use patient's DNA for further research purposes
Benefits of participating in testing	Availability of additional counseling and support services
Risks associated with genetic testing, including unexpected results	
Other available testing options	Acknowledgment of the right to refuse testing

Source: Abstracted from Andrews et al., 1995; Bove et al., 1997; Clayton et al., 1995, and Scanlon & Fibison, 1995.

the effects of medications (Bove et al., 1997). Other factors that may influence the ability to make an informed decision include who or how many among the health care team members provide the information, the amount of time allowed for communication of information and for reflection, whether the patient is alone or other family members are present, and the presence of concurrent illness or health problems. The setting in which informed consent is being obtained and insurance policy restrictions are other factors that may influence the informed-consent process.

The purpose of providing (i.e., *disclosing*) information to a patient or a family about genetic testing is to ensure that the consequences of testing are understood and that a truly informed decision can be made. The risks and benefits of testing should be explained fully, in a way that the specific patient or family is best able to understand. At the most basic level, this means that the information provided by nurses should be adequate for decision making, offered in a clear and understandable form, and given or repeated over time. During the process, patients should come to understand that they have a right to accept or to decline testing, without concern about the way they will be medically or emotionally treated. Throughout the process, the nurse must be aware of his or her own beliefs about the genetic testing and information gathered and ensure that she or he provides unbiased information and explanations.

Nurses can ensure that people considering genetic testing are aware of several other important issues during the informed-consent process. One such issue is ownership of samples. The nurse should inform the family that laboratories offering genetic testing may store and use samples for possible later research and testing. The nurse must explain that written consent is necessary to permit this future testing. Genetic testing might also discover nonpaternity, and a statement about this issue is usually incorporated into the informed-consent form for genetic testing. How such a situation might be handled is generally determined by the health care team that is offering the genetic testing and is discussed with the patient and family prior to testing.

Information obtained from genetic test results may have implications for other family members. The nurse who understands the importance of confidentiality and privacy will be better able to support patients' wishes. Before testing takes place, the nurse must know to whom the patient wants test results to be released. Does the person being tested wish to have other relatives informed? Disclosure of genetic testing information to insurance companies or employers may have untoward consequences, including the potential for discrimination. The nurse and health care team should review these issues before and during the testing process (Bove et al., 1997; Marteau, 1992; Scanlon & Fibison, 1995). Chapter 9 provides a more in-depth discussion of these ethical issues for nurses.

Genetic Screening

Both the potential risks and the potential benefits of genetic screening should be discussed during the informed-consent process. Such explanation is of great help to patients in decision making. Prenatal screening provides a good example.

Maternal serum alpha-fetoprotein screening, also called AFP profile screening, can give women and couples a wealth of important information. AFP profile screening involves the measurement of three substances made by the fetus, for the identification of pregnancies at increased risk for open neural-tube defects such as spina bifida and for chromosomal abnormalities such as Down syndrome. Screening that shows no increased risk can provide reassurance and enhance a developing relationship between a woman and the expected baby. When screening identifies a fetal abnormality, women can more readily consider and accept prenatal or early infant therapy, prepare for bearing and rearing a child with a disability, or consider terminating the pregnancy (ACOG, 1991).

The same prenatal screening, however, could increase anxiety, place excessive responsibility, blame, and guilt on a woman, interfere with parent-infant bonding, and disrupt the relationships among a woman, family members, and the community.

For many patients, deciding to have a screening test or undergoing the screening process elicits a mix of negative and positive reactions. Informed-consent explanations should help each patient to balance the risks and benefits of screening (Gates, 1993).

Carrier Testing

Like genetic testing and genetic screening, carrier testing has a range of risks and benefits, and again, an explanation of these potential risks and benefits for a specific individual can be the responsibility of the nurse (Fernbach & Thompson, 1992; Wilfond & Fost, 1990). The possibility of emotional distress, coercion, discrimination, changes in self-concept, and the use of genetic information for non-medical purposes are among the potential risks of carrier testing that nurses should explore with patients before they reach a decision to be tested (Marteau, 1992; Scanlon & Fibison, 1995). These factors tend to be present regardless of the condition that is the subject of the test. Carrier testing, however, can have risks and benefits that are both universal and specific for a condition or an individual. Cystic fibrosis (CF) serves as an example.

Carrier testing for CF is available for people with a positive family history. When people who could be offered CF carrier testing are identified by family history, nurses can tell them of the possible benefits. Test results can provide relief from worry. They can also help the patient to conduct informed reproductive planning and to make informed health care decisions, regardless of whether the test shows that the patient is a carrier. Some patients, of course, see as a burden rather than a benefit the opportunity to conduct informed reproductive planning or to make informed decisions about health care or relationships. Some women, for example, may not wish to know that they have passed on to a child a gene for a condition for which there is no specific treatment. The pathway to a decision is often not clear. The nurse, by providing information about the process and the range of consequences in a judgment-free manner, can help the patient to reach the most appropriate individual decision (Williams & Lea, 1995).

Presymptomatic Testing

Nurses may be the first health care providers to be approached by patients seeking information about presymptomatic testing for conditions such as Huntington's disease. Direct gene testing for presymptomatic diagnosis of many conditions, including Huntington's disease, is now available. For individuals at risk owing to a family history, such testing presents special concerns when a test proves diagnostic for a chronic degenerative condition for which there is no cure. Nurses can participate effectively in discussions with patients about such testing by knowing about current recommendations. For example, currently it is recommended that individuals with a positive family history of Huntington's disease who are interested in presymptomatic testing be referred to specialty centers where they can receive extensive education, counseling, and support before, during, and after testing (Huntington's Disease Society of America, 1989 and 1994; Motulsky, 1994).

Genetic Testing for Research Purposes

Some genetic testing takes place in clinical research settings. Genetic research testing may be concerned with sequencing genes, learning how genes function, or discovering treatments for genetic conditions. Nurses involved in offering genetic testing in this setting must consider special issues related to informed consent and the protection of patients and families involved in research studies. Questions for consideration are outlined and addressed by the Alliance of Genetic Support Groups (Alliance, 1997), a voluntary agency working in collaboration with a variety of professional and consumer organizations to develop informed-consent guidelines for research involving genetic testing (Figure 4.3). This brochure on informed consent, which outlines questions for patients and families to consider, may be useful for nurses who are preparing families for decisions regarding genetic testing and research participation.

Over the next few years, nurses will become increasingly involved in discussing and offering predisposition testing for such conditions as colon and breast cancer. This testing must be accompanied by extensive education and counseling with regard to its risks, benefits, and limitations, and may be carried out in collaboration with genetics specialty and research centers. Nursing issues in predisposition genetic testing are explored more fully in Chapter 8, and the ethical issues related to this testing are reviewed in Chapter 9.

FIGURE 4.3

Informed Consent: Participating in Genetic Research Studies

What is genetic research?

There are many different types of genetic research. Genetic research may be about finding genes (mapping), learning how genes work, or about treating or curing genetic conditions. To map a gene is to find its specific location, or a marker close by in your genetic code. Once a gene has been mapped, it is much easier to study the genetic basis of a condition. Scientists have to find out what the gene does (its function). They can then begin to develop tests for the gene and come closer to developing treatments and cures. Each genetic research project will have its own specific aims. These should be explained to you before you decide to participate in a project.

Genetic research projects often take place in hospitals or university medical centers. Before a project begins, it is usually reviewed by a committee called an Institutional Review Board or IRB. An IRB includes both scientists and nonscientists, such as clergy, community representatives, social workers, ethicists, lawyers, and nurses. The purpose of the IRB review is to assure that the interests of individuals participating in the research are well protected. However, the decision to participate is yours to make, and the informed consent process that precedes each study is designed to help you make your choice as freely as possible.

What is informed consent?

When researchers seek your consent, they are asking for your voluntary agreement to take part in a test, procedure, or research study. Informed consent means more than signing a written or printed consent form. This brochure is about informed consent for genetic research studies, and it also can help with informed consent for genetic tests and procedures.

To be informed, you need to know about benefits and risks of the research and how it may affect you, your family, and society. The research team should give you the information you need to make your own decision. Sometimes it is a good idea to talk about your decision with someone else not involved in the project. This person can be a family member, a good friend, or anyone you trust and respect and with whom you feel comfortable in discussing personal matters. Be sure to get the name and telephone number of a member of the research team you can contact in case you have any questions. In addition, the telephone numbers of other resources that you may wish to contact are listed in the back of this brochure.

Source: Reprinted with permission from the Alliance of Genetic Support Groups.

If you think you do not have enough information to make an informed decision, or there is something you do not understand, ask questions. Keep asking questions until you do understand the project. Only when you are sure that you know what the research involves and how it affects you and your family should you give your consent. Make certain you carefully read any document given to you for your signature. Also, you should be given a copy of that form for your personal records.

Who benefits from genetic research?

Individuals and families participate in genetic research for many reasons. They may simply be interested in working with scientists to advance the understanding of a genetic disorder, without expecting immediate benefit to themselves. Other studies may offer the possibility of developing clinical tests or clues to treatments that can directly benefit the participants and their families. Sometimes, participating in a study can put you in contact with specialists and specialized care that may be otherwise inaccessible. A critical part of the informed consent process is a clear explanation to you by the research team of the purposes of a study and the benefits that you and your family might expect to gain from participating.

What are some of the risks of participating in genetic research?

Genetic information about you or other family members may be discovered during the research project. Genetic information about you may indirectly provide information about your entire family. One of the first steps in most genetic research studies is to draw a family tree (a pedigree) that also contains some medical information. Family circumstances such as paternity and adoption may be revealed. Consider discussing with relatives how you as a family would feel about knowing these things or allowing the information to be given to others. Decide for yourself whether or not you want this new information. It is possible that even a nonparticipating family member could somehow be made aware of certain information by a researcher or, more likely, by another family member in general conversation.

Another outcome of participating in a genetic research study is that you may have the opportunity to learn information about your own genetic status, such as your risk of developing symptoms of a genetic disorder. You would need to decide whether or not you want this information for yourself or made known to other family members.

Some family members do not want to participate in research or know about certain information that could be found during the research project. Even if they do not want to participate in the research project, this genetic risk information could also apply to them. If this is the case, it is important that these issues be discussed with the researchers and a plan be worked out.

Information learned about you and your family, through your participation in genetic research, can become known to persons other than the research team. While researchers may try to protect you, no one can absolutely guarantee that at some point other researchers, insurance companies, employers, or other people will not get this information. You can ask for assurance that this information will not be put in your medical record. However, your insurance company will learn about it if you file a claim for any costs associated with this research project. Find out if your state has a law against discrimination based on genetic information.

There may be health-care costs to you for some types of genetic research. Your health insurance company may be unwilling to reimburse you for these costs. Inquire if your insurance company will pay for services related to the research project or experimental drugs. The informed consent agreement given to you by the research team should clearly spell out any costs which you may eventually have to pay.

Researchers will find it helpful to know why you or family members do not want to participate in their study, if this is the case. If you elect not to participate, your medical care and access to genetic counseling will not be affected. Ultimately, the decision as to whether or not to participate in the research study and whether to contact family members is yours.

Questions to Ask

The following is a list of questions that you might want to ask before deciding whether to participate in a genetic research study.

- *General Information*
  What is the purpose of the study? What are the names of the investigators? Who would be my contact person (and what is his or her phone number)? What agency is funding the research?

- *Benefits of Participating in Genetic Research*
  What are the benefits of participating in this research? For myself or family members? For others?

- *Risks of Participating in Genetic Research*
  What are the general risks of participating in this research? What physical risks may exist (beyond a blood sample)? What are some of the personal issues that could cause harm to me or my family (e.g., anxiety, stigma, discrimination, unpredicted disclosure of information)?

- *Treatment Issues*
Will treatment be provided if unexpected problems arise while I am participating in the study? Who will pay for this treatment?

- *Support & Special Services*
May I bring a friend or family member to help me, either while deciding to participate or while participating? Will special services be available for me if I need them (e.g., interpreters, braille, child care)?

- *Costs and Reimbursement*
How will costs associated with participation in this research be handled? Is there compensation for the time involved? Will the costs associated with travel/child care/special services be reimbursed? What additional health care costs may be associated with participation (e.g., will anything such as imaging scans and blood tests be billed to me or my insurance)?

- *Storage of Genetic Information*
What will happen to the stored DNA sample or any of my genetic information after this project is completed? What will happen if I decide to withdraw from this project? If this research plan changes in the future, if additional steps are added, or if new findings emerge, will I be notified and asked to sign another consent form? Will any of my genetic information be distributed (e.g., to pharmaceutical or biotechnology companies, genetic laboratories, or government agencies)?

- *Involvement of Other Family Members*
What happens if you need to have other family members involved in the study? How will they be contacted and by whom? What will happen to cells, DNA, or personal genetic information if they choose not to participate at all or withdraw from the study?

- *Study Results & Confidentiality Issues*
What will happen to the results of my tests from the study? Will I receive them? If so, how? May I choose not to receive the results (can I change my mind)? Who will get them (will they be put anywhere except in the research records)? How will confidentiality of the records, including photographs, be maintained?

- *Communication and Follow-Up*
How will the results of the research project be communicated to participants? If genetic services, tests, or treatments are developed from this research, how will I be told about their availability? How will I be informed if you publish information about me and my family? What happens if I do not participate in the research?

ADDITIONAL RESOURCES

Alliance of Genetic Support Groups
1-800-336-GENE
(301) 652-5553

Ethical, Legal, and Social Implications Branch
The National Center for Human Genome Research
National Institutes of Health
(301) 402-0911

National Society of Genetic Counselors
(215) 872-7608

Office of the Director
Office for Protection from Research Risks
National Institutes of Health
(301) 496-7005

NURSING PARTICIPATION IN MANAGEMENT OF PATIENTS WITH GENETIC CONDITIONS

Direct Nursing Care

The care of patients who have genetic conditions or who might develop such conditions will be of increasing importance to nurses. Regardless of the practice setting, the inclusion of actual or potential genetic concerns in nursing assessment, problem identification, planning, and interventions is integral to nursing care. Nurses will be better able to provide comprehensive, coordinated care if they have knowledge of the genetic contribution to such conditions as CF, Marfan syndrome, and familial cancers. Nurses must remain informed about the possibly changing medical, psychosocial, and counseling needs of patients who have these and other genetic conditions.

Criteria for nursing responsibilities and competencies (scope of practice) for basic-level and advanced nursing practice have been set forth by ISONG. Standards of genetic nursing practice are also being developed. These standards will provide the basis for further development of nursing responsibilities that incorporate genetic concepts and applications in the care of patients, families, and communities (Prows, personal communication, 1997).

Gene Therapy

Nursing care of people with genetic conditions may include participation in new and complex treatments such as gene therapy. The overall goal of gene therapy is to improve patient health by correcting the gene mutation in affected cells. For this goal to be achieved, normal genes must be delivered to the appropriate tissues. Gene therapy, therefore, involves gene transfer; functional copies of the relevant gene are transferred to achieve correction of the gene mutation in an individual (Kelly, 1994; Lea, 1997). Current gene-therapy protocols are investigating the use of this technology for treating patients with incurable genetic conditions (Anderson, 1994a; Jenkins et al., 1994).

The first gene-therapy research protocol was undertaken in 1990 to treat a single-gene disorder, severe combined immunodeficiency. Today, more than two-thirds of the available gene-therapy protocols are related to cancer. Further research is being carried out in the treatment of a wide range of genetic disorders (Anderson, 1994a, 1995). Nurses might participate in gene therapy as direct care providers, educators, advocates, providers of genetic services, and research investigators. Specific responsibilities for each of these roles are outlined in Table 4.6.

Patients and families may ask the nurse about the potential adverse effects of gene therapy. To date, gene therapy trials have not revealed any severe toxicity. Vigilant follow-up care of people receiving such treatments is necessary, however, to identify and prevent side effects. Nurses who care for gene therapy patients must develop a plan for monitoring these patients (Jenkins et al., 1994; Pickler & Munro, 1995; Wheeler, 1995).

TABLE 4.6

Nursing Responsibilities in Caring for Patients and Families Undergoing Gene Therapy

DIRECT CAREGIVER

Provides anticipatory guidance

Assures informed decision making and consent

Develops treatment and management plans

Administers gene therapy

Observes patients for expected and unexpected side effects of treatment including psychosocial and emotional response

Participates in developing long-term follow-up plans

Assures coordination and collaboration of care with all health care providers involved in the patient and family care before, during, and after gene therapy

EDUCATOR

Serves as an information source to patients, families, and public

Provides relevant, accurate, and understandable information to patients in both written and verbal format

Assures that patient and family questions will be answered at all times

ADVOCATE

Assures privacy and confidentiality of genetic information

Protects against discrimination

Advocates for fair and equitable use of gene therapies for all populations

Promotes public understanding of somatic gene therapy

GENETICS SERVICES PROVIDER

Gathers relevant family history information

Identifies individuals and families in need of further genetic education and counseling

Assesses psychosocial, ethnocultural, and educational background

Participates in genetic counseling process

Provides psychosocial support in follow-up to genetic counseling

RESEARCH INVESTIGATOR

Participates in or conducts clinical research trials in gene therapy

Serves as a preceptor to other nurses

Develops research protocols that will address patient and family response and adaptation to genetic information including gene therapy

Source: Reprinted from D. H. Lea, Gene therapy: current and future implications for oncology nurses. Semin. Oncol. Nurs. 13(2):115–122, 1997. With permission from the author and W. B. Saunders Company, Philadelphia, PA.

Gene therapy, as a clinical treatment for genetic conditions, is in its infancy, and it is not known whether or when it will become more widely used. Many questions remain for patients who are candidates for gene therapy or who have undergone treatment. For instance, what are the long-term psychosocial and physical ramifications of gene therapy? How do patients and their families adapt to new genetic therapies? What educational methods and materials are best suited for providing information about gene therapy? How can nurses prepare to care for patients and families before, during, and after gene therapy? Nurses may have opportunities in their practice to participate in nursing research to help answer such questions (Wheeler, 1991).

CONCLUSION

Nursing care of patients and families with genetic conditions is broadening as a result of the Human Genome Project. Nurses will participate in collecting genetic information by family and medical history and physical assessments and will become more involved in reporting and recording information from these assessments. Additional nursing activities are offering to patients appropriate and supportive information and resources about genetic tests and genetic screening, including the potential risks, benefits, and limitations of such testing. Nurses support informed health choices when they assure fully informed consent prior to genetic testing. This necessitates that they take into account patients' educational, social, economic, ethnic, and emotional status throughout the process. Nurses will also become increasingly involved with direct care of patients with genetic conditions and, therefore, must be knowledgeable about the genetic contribution to a given condition and about a patient's changing medical, psychosocial, and counseling needs. Gene therapy is an emerging treatment for genetic conditions. Nurses should be prepared, throughout the treatment, to care for patients and families who are participating in clinical gene-therapy research and to ensure comprehensive follow-up.

SUMMARY POINTS

As a result of information derived from the Human Genome Project, nurses are expanding their practice activities.

Collecting, reporting, and recording genetic information
- Obtaining family and reproductive history
  - family medical conditions having a genetic component
  - consanguinity
  - history of miscarriages, birth defects, maternal conditions and medications
  - maternal age
  - adoption
- Racial and ethnic background: important to note to be able to offer appropriate carrier testing
- Social, cultural, and spiritual assessment: influences health beliefs and views on the significance of genetic disorders
- Physical assessment: presence of a genetic condition suggested by certain physical features

Offering genetic information and testing
- Genetic screening: application of genetic testing to specific populations independent of a family history of a disorder; differs from genetic diagnosis in that it merely identifies patients who are more likely to have a genetic condition
- Nursing considerations in genetic screening and testing
- Assurance of informed choice and consent for prenatal, carrier, presymptomatic, and genetic research testing: support and advocacy

Nursing care of patients with genetic conditions
- Direct care: comprehensive care that includes knowledge of genetic contributions to heritable conditions and integration of that knowledge into care
- Gene therapy: participation in caring for patients undergoing gene therapy, a treatment to improve patient health by correcting the gene mutation in affected cells; development of plans for nurses to follow patients for expected and unexpected side effects

QUESTIONS FOR CRITICAL THINKING

1. List three practice skills that you would like to develop to help you to identify genetic information, interpret genetic information to your patients, or coordinate care of patients who have genetic conditions.

2. What aspects of the informed consent process are you providing currently? What additional materials and methods might you use to assist your patients in making informed choices?

3. Consider your past family and community-service experience with people who have genetic conditions and disabilities. How might such experience influence the way in which you provide information during informed-consent consultation and the type of information that you would provide?

REFERENCES

Alliance of Genetic Support Groups (1997). Informed consent: participation in genetic research studies. Chevy Chase, MD: Alliance of Genetic Support Groups.

American College of Obstetrics and Gynecology (1987). *Antenatal diagnosis of genetic disorders* (Tech. Bull. No. 108).

American College of Obstetrics and Gynecology (1991, April). *Alpha-fetoprotein* (Tech. Bull. No. 154). The American College of Obstetricians and Gynecologists, 409 12th St., SW, Washington, DC.

American College of Obstetrics and Gynecology, Committee on Genetics (1995, November). *Screening for Tay-Sachs disease* (Comm. Opin. No. 162). The American College of Obstetricians and Gynecologists, 409 12th St., SW, Washington, DC.

American College of Obstetrics and Gynecology, Committee on Genetics (1996a, February). *Genetic screening for hemoglobinopathies* (Comm. Opin. No. 168). The American College of Obstetricians and Gynecologists, 409 12th St., SW, Washington, DC.

American College of Obstetrics and Gynecology (1996b, February). *Hemoglobinopathies in pregnancy* (Tech. Bull. No. 220). The American College of Obstetricians and Gynecologists, 409 12th St., SW, Washington, DC.

Anderson, W. F. (1994). Gene therapy for genetic diseases. *Hum. Gene Ther.* 5:282.

Anderson, W. F. (1995). Gene therapy. *Sci. Am.* 9:124–128.

Andrews, L. B., Fullarton, J. E., Holtzman, N. A., and Motulsky, A. G., (eds.) (1994). *Assessing genetic risks: implications for health and social policy.* Washington, DC: National Academy Press.

Bove, C. M., Fry, S. T., and MacDonald, D. J. (1997). Presymptomatic and predisposition testing: ethical and social considerations. *Semin. Oncol. Nurs.* 13(2):135–140.

Clayton, E. W., Steinberg, K. K., Khoury, M. J., et al. (1995) Consensus statement: informed consent for genetic research on stored tissue samples. *J.A.M.A.* 274(22):1786–1792.

Cohen, F. (1984). *Clinical genetics in nursing practice.* Philadelphia: Lippincott.

Cystic Fibrosis Foundation (CFF), 6931 Arlington Road, Bethesda, MD 20814. Telephone: 1-800-344-4823 or 301-951-4422; fax: 301-951-6378.

Fernbach, S. D., and Thomson, E. J. (1992). Molecular genetic technology in cystic fibrosis: implications for nursing practice. *J. Pediatr. Nurs.* 7(1):20–25.

Fibison, W. J. (1983). The nursing role in the delivery of genetic services. *Issues Health Care Women* 4:1–15.

Forsman, I. (1994). Evolution of the nursing role in genetics. *J. Obstet. Gynecol. Neonatal. Nurs.* 23(6):481–486.

Gates, E. A. (1993). The impact of prenatal genetic testing on quality of life in women. *Fetal Diagn. Ther. 8(suppl 1)*:236–243.

Harper, P. S. (1993). *Practical genetic counseling* (4th ed.). Oxford: Butterworth.

Harrison, M. R., Golbus, M. S., and Filly, R. A. (1991). *The unborn patient: prenatal diagnosis and treatment.* Philadelphia: Saunders.

Hoang, G., and Erickson, R. (1985). Cultural barriers to effective medical care among Indochinese patients. *Ann. Rev. Med.* 36:229–239.

Holtzman, N. A. (1994). Discovery, transfer, and diffusion of technologies for the detection of genetic disorders: policy implications. *Int. J. Technol. Assess. Health Care* 10(4):562–572.

Huntington's Disease Society of America (1989). *Guidelines for predictive testing for Huntington's disease.* New York: Huntington's Disease Society of America.

Huntington's Disease Society of America (1994). The marker. *Newsletter* 7(1):6–9.

Jenkins, J., et al. (1994). Gene therapy for cancer. *Cancer Nurse* 17:447–456.

Kelley, W. N. (1994). The impact of gene therapy of medicine and society. *Ann. N.Y. Acad. Sci.* 716:12–19.

Lea, D. H. (1995). Advances in human genetics and their impact on primary care and obstetrical practice. *Genet. Teratol.* 4(2):1–4.

Lea, D. H. (1997). Gene therapy: current and future implications for oncology nursing practice. *Semin. Oncol. Nurs.* 13(2):115–122.

Lea, D. H., Williams, J. K., and Tinley, S. T. (1994). Nursing and genetic health care. *J. Genet. Counsel.* 3(2):113–124.

Marteau, T. M. (1992). Psychological implications of genetic screening. *Birth Defects* 28(1):185–190.

Milunsky, A. (1992). *Heredity and your family's health.* Baltimore: Johns Hopkins University Press.

Motulsky, A. G. (1994). Invited editorial: predictive genetic diagnosis. *Am. J. Hum. Genet.* 55:603–605.

Pickler, R. H., and Munro, C. L. (1995). Gene therapy for inherited disorders. *J. Pediatr. Nurs.* 10(1):40–47.

Prows, Cynthia. *Personal communication.* Chair, Professional Practice Committee, ISONG, 1997.

Rimoin, D., Connor, J. M., and Pyeritz, R. E. (1996). *Emery-Rimoin's principles and practice of medical genetics* (3rd ed). New York: Churchill Livingstone.

Roden, D. M., et al. (1996). Multiple mechanisms in the long QT syndrome: current knowledge, gaps and future directions. *Circulation* 94:1996–2012.

Scanlon, C., and Fibison, W. (1995). *Managing genetic information: implications for nursing practice.* American Nurses Association, 600 Maryland Ave., SW, Ste. 100 West, Washington, DC.

Schwartz, P. J., Moss, A. J., Vincent, M. G., and Crampton, R. S. (1993). Diagnostic criteria for the long QT syndrome: an update. Circulation 88(2):82–84.

Sudden Arrhythmia Death Syndromes Foundation (1997). The long QT syndrome: an information pamphlet for patients, families and physicians. Salt Lake City: SADS Foundation.

Thomson, E. J. (1993). Reproductive genetic testing: implications for nursing education. *Fetal Diagn. Ther. 8(suppl 1)*:232–235.

University of Colorado Health Sciences Center, School of Nursing and School of Medicine, Genetics Unit (1988). *Genetic applications: a health perspective.* Lawrence, KS: Learner Managed Designs.

Vincent, M. G., Timothy, K. W., Leppert, M., and Keating, M. (1992). The spectrum of symptoms and QT intervals in carriers of the gene for the long-QT syndrome. *N. Engl. J. Med.* 327:846–852.

Wenger, F. (1991). *Knowing yourself culturally: influences on clinical practice.* Paper presented at the Conference on Ethnocultural Diversity in the 90s. Florida Department of Health and Rehabilitative Services, Orlando, FL, October 29.

Wheeler, V. S. (1991). Preparing nurses for clinical trials: the cancer center approach. *Semin. Oncol. Nurs.* 7:275–279.

Wheeler, V. S. (1995) (suppl.). Gene therapy: current strategies and future applications. *Oncol. Nurs. Forum* 22:20–26.

Wilfond, G. S. and Fost, N. (1990). The cystic fibrosis gene: medical and social implications for heterozygote detection. *J.A.M.A.* 263(20): 2777–2783.

Williams, J. K. (1986). Genetic counseling in pediatric nursing care. *Pediatr. Nurs.* 12(4):287–290.

Williams, J. K. (1992). Guest editorial: meeting the challenge. *J. Pediatr. Nurs.* 7(1):2.

Williams, J. K. (1995). Genetics and cystic fibrosis: a focus on carrier testing. *Pediatr. Nurs.* 21(5): 444–448.

Williams, J. K. (1996). Genetic issues for perinatal nurses. March of Dimes Nursing Modules. White Plains, NY: March of Dimes Birth Defects Foundation.

Williams, J. K., and Lea, D. H. (1995). Applying new genetic technologies: assessment and ethical considerations. *Nurse Pract.* 20(7):16–26.

Wright, L., Brown, A., and Davidson-Mundt, A. (1992). Newborn screening: the miracle and the challenge. *J. Pediatr. Nurs.* 7(1):26–42.

CASE STUDY 4.1

GENETIC HEALTH HISTORY ASSESSMENT

The nurse in a family practice is providing care to the S family: Mrs. S, who is 41 years old and 12 weeks pregnant; Mr. S, age 50, and their 10-year-old son. At the first prenatal visit of Mr. and Mrs. S, the nurse obtains a genetic health history. Mr. and Mrs. S are both of Italian descent and are from large families. Mrs. S reports that she has had three spontaneous miscarriages between her pregnancy with her son and her current pregnancy.* She is concerned about this history and confides in the nurse, "We had given up hope that we

Women who are 35 years or older have a higher chance of having pregnancies in which a chromosomal abnormality is present. Prenatal testing, such as amniocentesis, is therefore offered and discussed. Italian descent is important to note, because individuals of certain ethnic backgrounds are at risk for being carriers of a recessive gene for genetic conditions such as beta-thalassemia. Carrier testing can be offered to both parents and, if identified as carriers, they can be offered prenatal diagnosis. Although miscarriage is common, occurring in approximately 15 percent of first-trimester pregnancies, repeated miscarriages may have a genetic basis for which testing is available. In 5 percent of couples who experience repeated miscarriage, inherited chromosomal abnormalities are the cause. Testing of the parents' chromosomes is offered and discussed, and if one of the parents is identified as a carrier, the issue of risk to other family members is raised and additional information is made available (ACOG, 1987; Rimoin et al., 1996).

would ever get pregnant again, and we are so afraid to get our hopes up this time." Mr. and Mrs. S tell the nurse that they want to do everything possible to ensure that their pregnancy is a success, but they have mixed feelings about prenatal testing because they do not know what they would do if they received abnormal test results.

- As a result of your genetic health assessment, what potential genetic issues have you identified for the couple in this pregnancy?
- What information might you provide about prenatal care and testing?
- What support might you give to Mr. and Mrs. S?

Discussion

The nurse in this setting carries out family and medical-history assessments, serves as an educator, and provides supportive care. The nurse can explain to Mr. and Mrs. S that in taking the family and medical histories, she has discovered several issues regarding which further information and discussion might be helpful: Mrs. S's age, their Italian ancestry, and a history of repeated miscarriage. Additional information and testing, which can be provided by a physician and genetic specialists if needed, might help Mr. and Mrs. S achieve the successful pregnancy they desire. The nurse can provide supportive care to Mr. and Mrs. S by exploring with them the reasons for their concerns about the pregnancy and noting this in the record; identifying their specific questions in preparation for their consultation with the physician; and explaining that many couples are unsure about what they might do with information obtained from genetic screening and testing during pregnancy and that additional support (e.g., genetic counseling) is available to them in making their decision.

Case Study 4.2

ASSURING INFORMED CONSENT FOR GENETIC SCREENING TESTS

Mrs. K is a 33-year-old woman who is 8 weeks pregnant with her second child. During Mrs. K's office visit, the nurse tells the patient about routine prenatal testing, including the AFP profile screening test.* The nurse explains that the AFP profile can help to provide information about risk for two categories of birth defects: neural-tube defects and chromosomal abnormalities such as Down syndrome. The AFP test can also provide information about pregnancy outcome. Mrs. K asks the nurse, "Do I have to have this test done? I am not sure

The AFP profile test is a prenatal screening test done at between 16 and 18 weeks of pregnancy to measure alpha-fetoprotein, estriol, and human chorionic gonadotropin, three substances made by the fetus and the placenta. Some of these substances pass from the fetus into the mother's bloodstream, where they can be measured. The AFP profile is called a screening *test because it identifies a few women in the population who appear to be at higher risk for carrying a fetus with certain birth defects, such as spina bifida and Down syndrome. A positive test result does not necessarily mean that a baby will have one of these disorders, but it does encourage the offering of additional prenatal testing such as high-resolution ultrasound examination and amniocentesis, for further evaluation of the pregnancy. If a birth defect is discovered to be present in the fetus, counseling, including genetic counseling, is offered to discuss the diagnosis in detail and the options available to the couple (ACOG, 1991; Harrison et al., 1991; Chapter 10).*

whether I want this kind of information about my baby. I didn't have it in my last pregnancy, and I'm still not sure in this one."

- How might you respond to Mrs. K's concerns?
- What additional information might you discuss with Mrs. K?

Discussion

The nurse recognizes her role as having to ensure that Mrs. K receives additional information about AFP profile testing and feels free to make an informed choice about whether to pursue prenatal screening. The nurse explains to Mrs. K that this testing is available to all pregnant women and that the choice of whether to have the test is Mrs. K's and her husband's. Additional written information suitable to Mrs. K's educational background and answers to questions that the couple asks could be offered. The nurse could take this opportunity to explore Mrs. K's concerns about AFP testing. One possible issue is that Mrs. K and her partner disagree about whether to have AFP testing. In collaboration with the health care team, the nurse might suggest referral for genetic counseling, explaining that Mrs. K and her partner would have the opportunity to discuss in detail the risks, benefits, and limitations of AFP testing and then to make the decision that is right for them (ACOG, 1991; ACDG, 1993; Thomson, 1993).

CARRIER TESTING FOR CYSTIC FIBROSIS

Mr. and Mrs. N have a 3-year-old daughter who has been diagnosed with cystic fibrosis.* At the cystic fibrosis clinic, they inform the nurse that Mrs. N's sister is pregnant and wants to have carrier testing for cystic fibrosis. Mr. and Mrs. N also have two other children, ages 5 and 9, and they say that they would like to have the children tested for carrier status. They ask the nurse to explain what is involved in the testing process.

- What information about cystic fibrosis carrier testing would you discuss with the family?
- What are some of the benefits and risks of carrier testing for children that you might relate?
- What is your obligation to Mrs. N's sister, when she calls and asks for information about Mr. and Mrs. N's daughter's diagnosis of cystic fibrosis?

*Cystic fibrosis is a common autosomal recessive condition in Caucasians, occurring in approximately 1 in 2500 live-born babies. People with cystic fibrosis have chronic pulmonary disease, pancreatic exocrine insufficiency, and increased concentration of sweat chloride. The genetic abnormality for cystic fibrosis has been identified, and the location is known to be on chromosome 7. Identification has allowed for DNA diagnostic, carrier, and prenatal testing for many families. When family members are seeking information about their carrier status, information about the gene mutation(s) for cystic fibrosis present in their family is helpful. Treatment of cystic fibrosis is evolving as new genetic knowledge becomes available (Cystic Fibrosis Foundation; Rimoin, 1996; Williams, 1995; Williams & Lea, 1995).

Discussion

The nurse in this clinical situation answers Mr. and Mrs. N's questions and provides resources for more information and assistance. The nurse can inform Mr. and Mrs. N that carrier and prenatal diagnostic testing are available for cystic fibrosis. Medical information about the specific cystic fibrosis gene(s) present in their daughter will allow for more accurate information about Mrs. N's sister's carrier status. The nurse informs Mr. and Mrs. N that neither she (or he) nor the office can release any information about their family's test results to other family members or health institutions without the permission of Mr. and Mrs. N; the nurse provides Mr. and Mrs. N with a consent-to-release form.

The nurse tells Mr. and Mrs. N that a genetic specialist is available to speak to them about the decision to test their children. The nurse might explain that a genetic counselor or genetic nurse specialist can review the possible benefits of testing, including learning that the N's children are not carriers. The nurse also might review some of the drawbacks of carrier testing, particularly in children, including the possibility of emotional distress, changes in self-concept and potential changes in parenting methods. Finally, the nurse could also explain that the children are not old enough to participate fully in the informed-consent and testing process.

CASE STUDY 4.4

CARING FOR PATIENTS CONSIDERING OR PARTICIPATING IN NEW GENETIC THERAPIES

Ms. P, a 20-year-old woman, has received a diagnosis of cystic fibrosis. During an office visit, she tells the nurse that she has been talking with her specialist about gene therapy.* Her specialist suggested that she consider being a participant in a current research trial. Ms. P asks the nurse, "What do you think I should do?" She tells the nurse that she has a boyfriend and that she and her boyfriend are considering marriage and a family. They would like to obtain more information about the chances that their children will have cystic fibrosis. Ms. P tells the nurse that her boyfriend is especially eager to know more about the disease and wonders when he can obtain the best information.

- How would you respond to Ms. P's question about gene therapy?
- What information could you provide about resources that might help to

Treatment for cystic fibrosis is currently directed toward managing pulmonary function and infection and improving nutrition. Identifying the gene mutation(s) for cystic fibrosis gives hope that the basic physiological problems that cause the condition will soon be better understood on a molecular level, paving the way for successful pharmacological interventions to correct the gene abnormality. Gene-transfer therapy is now being investigated for cystic fibrosis (Pickler & Munro, 1995; Lea, 1997).

answer Ms. P's question about the chances that her children will be affected?

Discussion

The nurse could respond to Ms. P's question about gene therapy by explaining that there is indeed a possible treatment for patients who have cystic fibrosis that is available on a research basis. The nurse then suggests that the best way for Ms. P to decide whether to participate in a research study is to obtain more information about its risks, benefits, and limitations. This information can be reviewed with Ms. P by the health care team at the research facility. Information on current research may also be available through the Cystic Fibrosis Foundation (Cystic Fibrosis Foundation). The nurse encourages Ms. P to discuss her concerns and feelings about the research study with her specialist and with her boyfriend.

The nurse could provide information about reproductive choices and support to Ms. P and her partner, explaining carrier testing and prenatal diagnosis for cystic fibrosis and could suggest that this is something Ms. P and her boyfriend might consider together. Because Ms. P has cystic fibrosis, her chances for having children with this condition are higher than the average person's. The nurse offers Ms. P and her boyfriend a referral for genetic counseling and explains that the genetic counselor and other genetics specialists will provide them with the most current information and will further discuss their questions. In making these suggestions, the nurse must be aware of her own values, beliefs, and experiences regarding genetic conditions and disability and reproductive choices such as pregnancy termination and of how these preconceptions affect the way in which the nurse presents and discusses information with the patient.

DNA RESEARCH TESTING USING LINKAGE ANALYSIS: LONG QT SYNDROME

Mary R is a 15-year-old patient in a cardiac clinic. After fainting while competing at a track meet, her electrocardiogram (ECG) and history suggest long QT syndrome (LQTS),* an inherited abnormality of the heart's electrical system. In talking with Mary's mother, the nurse notes that Mary has a history of fainting. The mother also relates that Mary has two brothers and a niece who died suddenly at young ages from cardiac arrest. Mary's mother tells the nurse that she has two other children, a 12-year-old boy and a 9-year-old girl.

Direct DNA testing, available for several gene mutations known to cause LQTS, has been performed for Mary and her mother. However, the specific gene mutations for which the tests were conducted were not identified in Mary's family. It is thought that other, as yet unidentified, genes may cause LQTS. The cardiologist in the clinic therefore offered Mary's family

LQTS is an abnormality of the heart's electrical system related to heart muscle cell structures called ion channels. These defects predispose affected persons to ventricular tachycardia or fibrillation (arrhythmia), which leads to sudden loss of consciousness (syncope) and may cause sudden death, often associated with exercise. QT refers to an interval measured on the ECG. The syndrome may be inherited or acquired. Two modes of inheritance have been described: an autosomal-dominant form, called Romano-Ward *type, and an autosomal-recessive form called* Jervell, Lange-Nielsen *type. Acquired LQTS is most often the result of administration of certain medications.*

The diagnosis is commonly suspected or confirmed on the basis of the ECG, with specific evaluation of the QT interval. The principle abnormality in LQTS is a prolonged QT interval relative to the heart rate and the appearance of abnormal T waves. Beta-blocking medications are the main treatment for patients with LQTS (Schwartz et al., 1993). These medications have been shown to be helpful in approximately 90 percent of affected individuals. Diagnostic DNA testing for LQTS is available in specified laboratories and is expected to improve the ability to make a firm diagnosis (Roden et al., 1996; Vincent et al., 1992).

the opportunity to participate in a research study that will involve linkage analysis or obtaining blood samples from multiple family members to try to track the gene in Mary's family. Mary's mother informs the nurse that the cardiologist told the family that this testing might be helpful in diagnosing LQTS in other family members and might help to ensure appropriate preventive treatment. Mary's mother asks the nurse to explain the testing process.

- If you were the nurse, how would you describe linkage analysis?
- What might you say about the risks and benefits of this testing?
- What other resources could you make available to the family to help them decide whether to pursue research testing?

Discussion

In this setting, the nurse can assume a number of roles. As an educator, she could provide an explanation of linkage analysis. She would relate that linkage analysis is a genetic testing method that involves using genetic markers to track the altered gene in a family. Linkage analysis makes use of specific genetic markers located on chromosomes near the gene in question, and these can be used to identify family members who may have the gene for LQTS. This testing requires obtaining blood samples from family members known to have LQTS, as well as from those who do not.

Other family members must be informed about the testing and must give their permission if they wish to be tested.

The nurse could make a referral to a genetic specialist within the hospital. She can explain that the genetic specialist will be able to provide details about the risks, benefits, and limitations of linkage analysis and that this generally includes reviewing and signing an informed-consent document. Some aspects of research testing that will be discussed include the accuracy of test results, the ways in which the DNA will be used during and after the research study, and the people to whom the test results will be available. The family might be given the brochure "Informed Consent: Participation in Genetic Research Studies" prepared by the Alliance of Genetic Support Groups (1997), with the explanation that this brochure provides additional information about genetic research testing that may be helpful to family members as they discuss their decision about the testing (see Chapter 4 for further discussion).

Ensuring support and access to current genetic information are two other nursing roles in this situation. The nurse can explore current family and community supports with Mrs. R and can ensure that the family is aware of the LQTS support group called the *Sudden Arrhythmia Death Syndromes (SADS) Foundation* (1997), a parent organization that provides current information and support to families affected by LQTS.

What to Expect from Genetic Counseling

To care for another person, in the most significant sense, is to help him grow and actualize himself.   *Milton Mayeroff*

OBJECTIVE

Describe the elements of genetic counseling and nursing participation in this process.

RATIONALE

Nurses help to prepare patients for many health conditions. Primary nursing roles are to participate in genetic counseling and to support patients throughout the counseling process. Nurses also help patients to integrate new genetic information into daily life. To perform these roles, nurses must have knowledge and understanding of genetic counseling.

APPLICATION ACTIVITIES

- Identify and refer patients and families who may benefit from genetic counseling.
- Provide appropriate genetic information prior to, during, and in follow-up to genetic counseling.
- Help to gather relevant family and medical history.
- Collaborate with genetic specialists and provide support to patients and families throughout the genetic counseling process.
- Coordinate genetic health care.
- Identify relevant community and national support resources.

GENE
CARE
LINK

For a listing of additional genetic resources and services, go to http://www. jbpub.com/ clinical-genetics.

INTRODUCTION

*G**enetic counseling as a clinical genetic service grew out of a
need for professionals who could provide genetic information, education, and support to patients and families with current and future genetic health concerns. Genetic counseling was
first described as a discipline encompassing the knowledge of
human genetics; respect for the sensitivities, attitudes, and reactions of patients; and a desire to teach the truth to the fullest
extent (Cohen, 1984; Reed, 1980). Genetic counseling has since
evolved to mean a communication process by which individuals
and families come to learn and understand relevant aspects of
genetics, to make informed health decisions, and to receive support
in integrating personal and family genetic information into their
daily lives.*

*A formal definition of genetic counseling was developed in
1975 by the American Society of Human Genetics Ad Hoc Subcommittee on Genetic Counseling (Epstein, 1975). According to
this definition, genetic counseling is a communication process
that deals with the human problems associated with the occurrence, or the risk of occurrence, of a genetic disorder in a family.
This process involves an attempt by one or more appropriately
trained persons to help the individual or family to*

- *comprehend the medical facts, including the diagnosis, probable
  course of the disorder, and the available management;*
- *appreciate the way heredity contributes to the disorder and the risk
  of recurrence in specified relatives;*
- *understand the alternatives for dealing with the risk of recurrence;*
- *choose the course of action that seems to them appropriate in view of
  their risk, their family goals, and their ethical and religious standards, and to act in accordance with that decision; and*
- *in an affected family, to make the best possible adjustment to the disorder or to the risk of recurrence of that disorder (Epstein et al., 1975).*

Genetic counseling has traditionally been provided in regional genetics centers, medical centers, and community-based hospitals and in outreach and public health clinics. As genetic discoveries increase knowledge of genetics and health and as the number of genetic tests and treatments grows, genetic counseling has expanded into other clinical settings such as managed health care organizations, commercial facilities, and private practices. Primary care providers, including nurses, will therefore participate more in genetic counseling (Andrews et al., 1994; Forsman, 1994). Chapters 2 and 4 offer detailed information about the role of genetics in health care and the integration of genetics into nursing practice. This chapter provides an overview of the components of the genetic counseling process and addresses the participation of nurses in preparing and supporting patients and families throughout their counseling experience. Additional discussions of genetic counseling can be found in books and articles referenced at the end of the chapter.

COMPONENTS OF GENETIC COUNSELING

Genetic Specialists

Genetic counseling is provided by a team of specialists that includes medical geneticists, genetic counselors, or advanced-practice genetic nurse specialists. Genetic specialists have professional education in genetics that prepares them to provide specific genetic services. Table 5.1 summarizes the qualifications of genetic specialists and their roles in providing genetic counseling.

Team Approach

The team of genetics specialists collaborates to provide support, genetic evaluation, and information and resources to patients and

TABLE 5.1

Qualifications and Roles of Genetic Specialists

Clinical geneticist: a physician (M.D. or D.O.) who provides comprehensive diagnostic, management, and counseling services. Clinical geneticists come from a variety of disciplines, including pediatrics, internal medicine, obstetrics and gynecology, ophthalmology, and dentistry.

Ph.D. medical geneticist: a person with a doctoral degree who works in association with a medical specialist in a clinical genetics program.

Clinical molecular geneticist: a doctor (M.D. or Ph.D.) who performs molecular genetics testing analyses relevant to the diagnosis and management of genetic conditions. Clinical molecular geneticists might be clinical medical geneticists or individuals who hold a doctoral degree in molecular genetics or molecular biology.

Genetic counselor: a health professional who has a Master's degree and education and training appropriate to providing genetic-counseling services to individuals and families who are seeking information about genetic conditions or birth defects. Genetic counselors practice as members of the genetics team and come from such backgrounds as biology and other basic sciences, social work, and nursing.

Advanced-practice nurses in genetics: a nurse with a Master of Science or doctoral degree in nursing who practices as a member of a genetics team and provides evaluation, counseling, and educational services. Advanced-practice nurses come from nursing disciplines that include maternal and child health, oncology, neurology, hematology, and endocrinology.

Genetics nurse: a nurse who provides nursing care to patients and families with a specific genetic condition or need for a specific genetics service. Genetics nurses can be found in a diverse number of clinical settings specific to the genetic condition.

Source: Adapted from CORN, Guidelines for Genetics Service Delivery, 1997.

families. In some settings, a social worker with professional training in genetics participates as a team member. Working together, the specialists obtain and interpret complex family history information; evaluate and diagnose genetic conditions, interpret and discuss complicated genetic test results, support patients throughout the genetic counseling process, and offer resources for additional professional and family support. The patient and family participate as team members and decision makers throughout the counseling process.

Access and Referral to Genetic Counseling

People obtain genetic services in a range of ways. Most states provide federal or state funding for regional genetics services to ensure that all patients and families will have access to such services. Patients and families who have genetic concerns are usually referred for genetic counseling by primary care providers, including nurses. In some situations, families learn of genetic counseling services through consumer advocacy groups. These groups can help to identify which genetic counseling services are appropriate for the patient's condition or concern and can assist with locating such services. Patients and families sometimes make a direct request for genetic counseling. The genetics professional, in these circumstances, assists the patient in determining appropriate genetic counseling and coordinates this type of referral with the primary care provider or other caregiver.

Patients come to genetic counseling for a variety of reasons and at different stages of life. Some are seeking preconception or prenatal information; others are referred following the birth of a child with a birth defect or suspected genetic condition; still others seek information for themselves and their families because of a family history of a genetic condition. Regardless of the timing or setting, genetic counseling should be offered to all patients who have questions about genetics and their health. The nurse should consider referring for genetic counseling any patient in whose family a heritable condition exists and who asks, "Will I get it? What could cause this disorder to affect me, my child, or another relative? Will it happen again in other pregnancies or relatives? What are my reproductive and health options?" Nurses, in collaboration with other health care professionals, refer patients and families for other specific genetic concerns, which are described in Chapter 4.

RELEVANT ISSUES IN GENETIC COUNSELING

Nondirectiveness

Nondirective provision of information has evolved as a guiding principle of genetic counseling, out of respect for patients' rights to self-determination. *Nondirectiveness* means supporting patients' decisions that reflect their personal values, beliefs, and interests. Genetics specialists who provide counseling must make every attempt to respect patients' abilities to make autonomous decisions. The first step in providing nondirective counseling is recognizing one's values and how communication of genetic information may be influenced by them (Fine, 1993; Sorenson, 1993).

Nondirectiveness involves providing information about all facets of the genetic condition or concern, whether that be the risks, benefits, and limitations of a particular genetic test or the specific features, prognosis, and management of the genetic condition being considered. Circumstances surrounding the identification of and treatments for many genetic conditions are changing as more is learned about the contribution of genetics to disease. Improvements in life expectancy have been made for a number of genetic conditions such as sickle-cell anemia and cystic fibrosis, and advances in treatment such as gene therapy are occurring. The genetic specialist discusses genetic conditions and traits in a balanced manner, presenting these medical advances while maintaining respect for individuals and genetic differences. Further considerations of ethical issues and nondirectiveness are presented in Chapter 9.

Awareness of Social and Cultural Differences

Assessment of social, emotional, cultural, and spiritual beliefs, which is important as the nurse obtains, records, or reports genetic information (see Chapter 4), is also important during genetic counseling. Genetic specialists include in their counseling an assessment of social and cultural differences, noting how patients may value and interpret genetic information. Cultural and language barriers, family decision-making processes, concepts of health and disease, and the expectations of the genetic counselor and the institution are considered throughout the counseling process. Genetic specialists must be aware of their own cultural values and of how these may influence the manner in which they collect and discuss information. Recognizing these personal views, they must strive to convey genetic information during counseling in a way that demonstrates appreciation of individual choices and views (Fisher, 1995).

Confidentiality and Privacy

Confidentiality of genetic information and respect for privacy are essential ethics underlying all genetic counseling. Ethical issues related to confidentiality and privacy are discussed in greater detail in Chapter 9. All genetic specialists, including nurses who participate in the genetic-counseling process and who have access to genetic information, should honor the patient's desire for confidentiality. Information should be kept from family members, insurance companies, employers, and schools if that is the patient's desire, despite special circumstances that might make such confidentiality difficult. The genetics professional, for example, may explain the need to disclose genetic information to other family members who could experience significant harm if such information is not shared. The patient may have contrary views and may wish to keep this information from the family. Such differences can create a difficult situation for both patient and provider. Nonetheless, the genetics specialist helps to support the patient's wishes while explaining the potential benefit of this information to family members.

THE GENETIC-COUNSELING PROCESS

Genetic counseling is a process and may take place over an extended period. Genetic counseling may, therefore, entail more than one visit. The purpose is to allow patients and families to learn and understand genetic information, to gain support and guidance in decision making, and to receive comprehensive and coordinated care if they have specific genetic conditions or concerns. The components of the genetic counseling process are outlined in Table 5.2.

TABLE 5.2

The Genetic-Counseling Process

ASSESSMENT AND INFORMATION GATHERING

Reason for referral

Information gathering

Family history
Medical records
Relevant testing results
Social and emotional concerns

Identification of relevant cultural, educational, and economic factors

EVALUATION AND ANALYSIS OF DATA

Family history

Physical examination as needed

Further laboratory testing

Other relevant testing (ophthalmological, neurological)

INFORMATIVE COUNSELING: COMMUNICATION OF GENETIC INFORMATION

Natural history of condition

Inheritance pattern

Reproductive and family health options and issues

Testing options

Management issues

SUPPORTIVE COUNSELING

Eliciting family and individual questions and concerns

Identifying existing support systems

Providing emotional and social support

Referring for additional counseling and support as needed

FOLLOW-UP GENETIC COUNSELING

Further discussion of test results and diagnosis

Written summary to referring care providers and family

Coordination of care with primary care providers and other specialists

Genetic counseling may be offered at any point during the life cycle. Specific genetic-counseling issues are often related to the time in the life cycle that counseling is sought (Table 5.3). The following discussion of genetic counseling is an overview of the genetic-counseling process, with examples of counseling concerns and issues for prenatal, newborn, pediatric, adolescent, and adult patients; their families; and their health care providers. More detailed discussions of genetic counseling can be found in the texts referenced in this chapter.

TABLE 5.3

Examples of Genetic-Counseling Issues at Various Life Stages

PRENATAL PATIENTS

Understanding of screening versus diagnostic testing

Inplications of reproductive choices

Impact on maternal/parental-fetal bonding

Potential for anxiety and stress

Effects on partnership and family

NEWBORN PATIENTS

Potential for disruption of parent-newborn relationships

Parental guilt

Implications for siblings and other family members

Coordination and continuity of care

PEDIATRIC PATIENTS

Caring for children with complex medical needs

Coordination of care

Potential for alteration in parent-child relationships

Potential for social stigmatization

ADOLESCENT PATIENTS

Potential for decreased self-esteem

Potential for alteration in perception of family

Implications for lifestyle and reproduction

ADULT PATIENTS

Potential for ambiguity of test results

Identification of susceptibility without an existing treatment or cure

Effect on marriage, reproduction, parenting, and lifestyle

Impact on insurability and employability

Nurses can refer patients and families, collaborate with genetics specialists, and participate in genetic counseling by

- providing appropriate genetic information prior to, during, and in follow-up to genetic counseling;
- helping to gather relevant family and medical history information;
- offering support to patients and families throughout the genetic-counseling process; and
- coordinating genetic health care with relevant community and national support resources.

These collaborative activities help to ensure that all patients and families receive the most benefit from genetic counseling.

Precounseling and Information Gathering

The genetics service to which the nurse refers a patient or family for counseling will ask the nurse to provide background information for assessment. The genetics specialist needs to know the medical reason for referral, the patient's or family's reasons for seeking genetic counseling, and potential genetic-counseling issues. The nurse may refer a family with a new diagnosis of familial adenomatous polyposis to discuss the inheritance and implications for other family members. The family may have concerns about confidentiality and privacy. Using the nursing assessment, the genetics specialists will tailor the genetic counseling to respond to these concerns.

Other information provided by the nurse includes relevant test results and medical and developmental evaluations. The nurse must secure permission from the patient and, if needed, other family members, to obtain, review, and transfer medical records that document the genetic condition of concern. Medical genetic evaluation of more than one family member may be necessary in some situations to establish a genetic diagnosis. The nurse can prepare patients for this assessment by explaining that the medical information and evaluation are necessary to ensure that appropriate information and counseling (including risk interpretation) are provided.

The nurse will be asked to provide information about the social and emotional status of the patient and family. The genetics specialist will inquire before the visit about potential financial issues, so that appropriate financial assistance can be initiated. The specialist will want to know about the coping skills of a family that has recently learned of a genetic diagnosis. He or she will want to be aware of the types of genetic information being sought. The nurse helps to identify cultural issues that may influence how information is provided and by whom. For patients who are hearing impaired, for example, an interpreter's services may have to be arranged in advance. The genetics professional, after exploring these issues with the nurse, records the information in the patient record.

Evaluation and Analysis of Data

Family History

Before counseling begins, the nurse can discuss with the patient and family the type and nature of family-history information that will be collected during genetic counseling. Family-history interpretation, an essential component of genetic counseling, is used to help determine the pattern of inheritance of a genetic condition and the chance of inheritance for other family members (e.g., children, siblings, cousins). The family history, as discussed in detail in Chapter 3, is recorded as a genetic pedigree and provides a graphic representation of the family medical history (Bennett et al., 1995). The medical elements included in a genetic family pedigree are outlined in Table 5.4. Family-history collection and analysis is comprehensive and focuses on obtaining information that may be relevant to the specific genetic concern in question. Although targeted to each genetic-counseling situation, such analysis always includes assessment for any other potentially inherited conditions for which testing and preventive and treatment measures might be offered.

Prenatal Patients

In a prenatal setting, genetic counseling is often necessary for pregnant women who have an abnormal alpha-fetoprotein screening test result. If the test indicates an increased risk for a neural-tube defect,

TABLE 5.4

Medical Elements of the Genetic Family Pedigree

All births	Other illnesses, including cancer, heart disease, diabetes
All deaths, ages at, and causes of death	
Inherited and familial disorders	Age at diagnosis
Confirmation by medical record or pathology report of all genetic and medical diagnoses	Miscarriages, stillbirths, elective pregnancy terminations
	Birth defects

Source: Bennett et al., 1995; Peters and Stopfer, 1996.

the genetic specialist, when collecting family history from the nurse and the patient, would ask whether any family members had been born with a neural-tube defect such as spina bifida. The information would be included by the genetics specialist in the analysis of risk to the fetus.

Newborn Patients

When, for example, a baby is born with features suggestive of Down syndrome, the genetic specialist reviews the family history with the couple, asking whether any other family members have Down syndrome. This information will help the specialist to determine whether Down syndrome is inherited in that family. Maternal age is also noted, because women over age 35 are at higher risk for giving birth to a baby with Down syndrome. The genetic specialist collects and analyzes this information and the chromosomal tests for Down syndrome in the infant to help counsel the family about the cause and chance of recurrence of Down syndrome.

Pediatric Patients

For a child with a cleft lip and palate, the family-history analysis would include determining whether other family members have cleft lip and palate or other abnormalities, such as congenital heart defects or vision, hearing, joint, or skin problems. This information will help the genetic specialist to determine whether cleft lip and palate has occurred as an isolated condition in the child or is a part of a genetic syndrome that could affect other family members.

Adolescent Patients

In referring for genetic counseling a tall, thin adolescent who has a heart murmur, the nurse should ensure that a complete family history is available, for a possible diagnosis of Marfan syndrome. The nurse should ask about the presence of visual, skeletal, and cardiac problems in other family members. In this situation, the family history is important for diagnosis, as well as for identifying other family members who might be at increased risk for Marfan syndrome.

Adult Patients

Families with a history of cancer often request an analysis to determine the nature and degree of their risk. The genetic specialist conducting a family-history analysis to address a concern about breast cancer, for example, would inquire about the number of affected and unaffected relatives on both the maternal and paternal sides of a family, the age at diagnosis, and the presence of other cancers in the family, to determine whether the cancer is familial.

Physical Examination

A physical examination performed by a medical geneticist may be needed to look for specific clinical features that are diagnostic of a genetic condition. The examination also helps to identify whether additional laboratory work, such as chromosomal analysis or DNA testing, would help to clarify a genetic diagnosis. The detailed physical examination generally involves assessment of all body systems, with a focus on specific physical characteristics considered for diagnosis. Before the genetic-counseling visit, the genetic specialist can provide the nurse with information about the nature and focus of the physical examination, so that the nurse can discuss these diagnostic evaluations and their purpose with patients and families, in preparation for the counseling.

Prenatal Patients

In prenatal genetic counseling, the physical examination usually involves a detailed evaluation of the fetus. During a prenatal genetic-counseling session, if there is a question of a neural-tube defect in the fetus, for example, the genetic specialist or other medical specialists perform a high-level ultrasound examination of the fetus, paying special attention to the spine, head shape, and size of structures within the brain. All physical features of the fetus are evaluated for signs of other abnormalities that might suggest the presence of a genetic syndrome. In this setting, the physical examination is important for diagnosis and for discussion of options, such as further evaluation by amniocentesis.

Newborn Patients

The physical examination of an infant suspected of having Down syndrome involves a careful inspection of physical features. The shape of the head and the shape, size, and location of the eyes, ears, nose, and mouth are considered. The baby's hands are examined for the presence of a single palmar crease, called a *simian crease,* and for curvature of both little fingers. A careful cardiac examination is performed, because 40 percent or more of patients with Down syndrome have a congenital heart defect. The examination provides information that will be analyzed in preparation for diagnostic chromosomal analysis.

Pediatric Patients

A child with cleft lip and palate is given a detailed physical examination to determine the presence of other physical features that might suggest a genetic syndrome. The medical geneticist pays special attention to such clinical features as the number and placement of teeth, facial structure, skin and hair textures, hearing, vision, and the cardiovascular system. A brief physical examination of the parents, siblings, and other relatives for evidence of cleft lip or palate and for other features that might suggest the presence of a familial genetic syndrome may be included in the evaluation, in preparation for making a diagnosis.

Adolescent Patients

The physical examination is central to the diagnosis of Marfan syndrome. The medical geneticist pays special attention to growth parameters, such as height, lower- to upper-body ratios, and arm span. Hand and finger shape and size are noted. Other skeletal features, such as the presence or absence of scoliosis and chest shape and size, are examined and recorded. An echocardiogram to evaluate the heart and aortic root diameter, and a detailed eye examination to look for the presence of lens dislocation may be recommended, depending on the results of the physical examination and family-history analysis.

Adult Patients

If a patient reports a family history of early-onset breast or other cancers and medical records document the diagnosis, the medical specialist performs a detailed physical examination for evidence of cancer, such as tumors or skin or eye findings. The examination provides a basis for development of a plan for further discussion of diagnostic or presymptomatic testing for cancer (Peters and Stopfer, 1996).

Communication of Genetic Information

The genetics team reviews the family and medical histories and results of the physical examination before beginning genetic counseling with the patient and family. The genetics specialists then meet with the patient and family to discuss their findings. When the information gathered confirms the presence of a genetic condition in a family, the specialist conveys the natural history of the condition, the pattern of inheritance, and the implications of the genetic condition for reproductive and general health and, when appropriate, offers relevant testing and management options.

Prenatal Patients

When a prenatal diagnosis of spina bifida is confirmed, the genetics team explains to the woman, couple, and family that spina bifida is a neural-tube defect that usually occurs as a result of multifactorial inheritance. *Multifactorial* means that environmental and inherited factors from each parent work together during fetal development and interrupt completion of spinal development. If diagnostic tests have identified the cause of spina bifida in the fetus as chromosomal, the team explains the type and nature of the chromosomal abnormality. The genetics team also discusses the medical implications for a baby born with spina bifida and offers the patient or couple the opportunity to meet with other specialists, such as pediatric neurologists and neurosurgeons, to discuss prognosis and treatment options. The genetics team provides information about other community resources such as parent support groups and social services. The team also informs the woman and her partner of options for pregnancy termination and of the alternative of continuing the pregnancy to term with the option of giving up the child for adoption.

Newborn Patients

The genetics team provides comprehensive information to women, couples, and families who have a newborn with a confirmed diagnosis of Down syndrome. The geneticist explains that the diagnosis of Down syndrome is confirmed when an extra number 21 chromosome is observed on chromosomal analysis. If the chromosome analysis reveals that the Down syndrome is due to a complete extra number 21 chromosome that occurred as a result of nondisjunction, the genetics specialist notes that this is a random event. If, on the other hand, the analysis reveals that the number 21 chromosome was inherited as a result of an unbalanced translocation, the genetics specialist explains the heredity of Down syndrome in some families and discusses further testing of the parents and other relatives to learn more about how Down syndrome may be inherited in that particular family.

The family is given information about the natural history of Down syndrome, and the special medical, educational, and social needs of the child and the family. The genetics team provides written information about Down syndrome, and identifies local and community resources for the family, recognizing that families are not always immediately ready for information and resources but will know when the time is right for them.

Pediatric Patients

In counseling a family with a child with cleft lip and palate, the genetics team focuses on the cause of cleft lip and palate in the family, the implications for other family members, and the child's medical-management needs. If the child's cleft lip and palate are a part of a genetic syndrome, then the genetics specialist describes the syndrome, associated health problems, and the need for ongoing follow-up. The genetics specialist also talks with the family about the inheritance of the syndrome and the importance of informing other family members about the condition so that they can be offered genetic evaluation and counseling. In this counseling situation, as in many others, such additional evaluations as cardiac, ophthalmological, or hearing tests may be recommended. With the family and the primary care provider, the genetics team develops a coordinated plan for further testing, management, and ongoing care.

Adolescent Patients

Providing genetic counseling to an adolescent with a new diagnosis of Marfan syndrome involves discussing with the patient and his or her family the clinical findings of the physical examination and explaining how these fit together to create the diagnosis. The medical geneticist and genetic counselor or nurse specialist explain the underlying defect in a connective-tissue protein, determined by a gene on chromosome number 15. They discuss the inheritance pattern and review treatment and management options. The medical geneticist and counselor also review the potential impact of the diagnosis on life style, explaining the need for the patient to avoid strenuous exercise to minimize stress to the aorta.

Adult Patients

For patients with a confirmed family history of breast or other hereditary cancer, genetic counseling focuses on discussion of the inheritance of cancer, presymptomatic and diagnostic testing and treatment options, and the implications for family members. A team of specialists, which might include oncologists, geneticists, genetic counselors, genetics nurse specialists, and social workers, may be involved throughout the genetic-counseling process in talking with the patient and family about the risks, benefits, and limitations of further testing and treatment. Genetics professionals provide information in understandable language and in a manner appropriate to the patient's and family's educational, social, and cultural background. Visual educational materials are used when discussing the condition and its inheritance pattern.

Genetic-counseling sessions may last for 1 hour or more, depending on the setting and the patient's and family's needs. Support and follow-up are always offered, and the fact that genetic information may take time to understand and to assimilate is explained. The team assures the patient and family of the team members' ongoing availability. (Further details of genetic counseling and cancer-risk assessment are provided in Chapter 8.)

Supportive Counseling

By confirming a genetic diagnosis and providing genetic information, genetics specialists might prompt patients to share additional personal and family issues. The genetics team provides supportive counseling throughout the counseling session and makes every effort to elicit individual and family concerns. In all genetic-counseling settings, principles of active listening are used to interpret patient concerns and emotions, seek and provide feedback, and demonstrate understanding of those concerns. If needed, the genetics professional may suggest referral for additional social and emotional support.

The genetics professional discusses pertinent patient concerns and needs with the nurse and primary health care team so that they can provide additional support and guidance to the patient. In each of the situations described here, the genetics specialist discusses identified patient concerns and needs with the primary care provider and the nurse so that they can continue to provide ongoing support and guidance to the patient and family.

Prenatal Patients

Confirmation of a diagnosis of an open neural-tube defect in a fetus is often devastating for women, their partners, and their families, and may precipitate feelings of loss and guilt. Each person involved may experience grief differently, and at different times. Genetics specialists validate these feelings and discuss with the patient the importance of acknowledging and sharing feelings. The patient is given the opportunity to ask questions and to discuss the options of pregnancy continuation, pregnancy termination, and adoption. Genetics specialists offer support to the woman, her partner, and her family, letting them know that there are no right or wrong decisions and that the "right" decision in this situation is whatever choice best suits that woman and her family at the time in accordance with their personal philosophy and values. Genetics specialists offer additional psychosocial support, including, if possible, an opportunity for the patient to speak with other patients who have faced similar decisions. They also make available appropriate written materials about neural-tube defects and decision making in this situation (Minnick et al., 1996).

Newborn Patients

The family with a newborn newly diagnosed with Down syndrome will need support to understand the many developmental and health issues for the affected child. Supportive genetic counseling in this situation consists of responding to the family's questions as they arise. Some families experience a sense of loss and grief because they did not have the baby they expected, whereas others, because of their life experiences with family members who have Down syndrome or other disabilities, may adapt to the diagnosis in a different way. Genetics team members might explain to the family that when parents learn that their baby has a chronic condition for which lifetime support is needed, the parents may experience chronic sorrow. Chronic sorrow has been observed in families responding to a diagnosis of Down syndrome and other chronic illnesses in a child and occurs throughout the lifetime of the child. Sorrow may recur during times of stress or changes in the child's condition (Clubb, 1991). The genetic specialist helps the family to understand this process, assesses the family's emotional and social needs and supports, and provides appropriate information, including community and national resources (Clubb, 1991; Cohen, 1984).

Pediatric Patients

Supportive counseling for the family whose child has cleft lip and palate, whether an isolated disorder or a syndrome, includes providing both support for the family and information about medical care for a child with complex health needs. The family is acknowledged as an essential team member in all management and treatment plans for the child. The genetics team, often in collaboration with a craniofacial clinical team, explores with the family the medical, social, and educational resources that are available and develops a plan that will be coordinated by the primary care provider and nurse and that ensures continuity of care. This plan includes offering parents support through local and national genetics support organizations.

Adolescent Patients

The adolescent with a new diagnosis of Marfan syndrome is included from the outset in discussion and counseling about the condition. Supportive counseling is ongoing, and may be needed beyond the genetic-counseling encounter. The genetics specialists assess developmental and psychosocial issues and encourage the patient to share questions and concerns. A teenage girl receiving a diagnosis of Marfan syndrome, for example, may experience feelings of decreased self-esteem. At the same time, her parents may be struggling with alterations in their perception of the family as a whole or with feelings of guilt if one of the parents also has Marfan syndrome. The genetics specialists identify these concerns and explain to the family that these feelings are not unusual among families in which a new genetic diagnosis has been made. Supportive information available through the National Marfan Foundation, including information about peer support and a telephone help line, may be offered, and additional supportive counseling may also be offered as needed.

Adult Patients

Complex issues must be considered by patients and families undergoing genetic counseling and presymptomatic and diagnostic testing for breast and other cancers. Supportive counseling is offered by the entire team of specialists throughout the risk-assessment and genetic-counseling process. A coordinated and collaborative approach involving multiple specialists is currently recommended (Biesecker, 1997) and is discussed in greater detail in Chapter 8.

Follow-Up Genetic Counseling

Follow-up genetic counseling is always offered to patients and families. A written summary of the genetic evaluation and counseling session(s) is prepared in follow-up to genetic counseling and, with the patient's permission, sent to the primary care provider as well as all other providers and participants in the patient's care, as identified by the family. The report summarizes the results of any physical examination, discusses the inheritance and associated recurrence risks for the patient and family, presents reproductive options, and makes recommendations for further testing and management. A separate letter is often written to the patient and family, in language tailored to each patient's educational and information needs. A copy of these reports, maintained in the medical genetics records, can, with the patient's permission, be sent to other care providers as needed.

Follow-up genetic counseling is available to patients and families who may need more time to understand and discuss the specifics of a genetic test or diagnosis or who might wish to review reproductive options again later when pregnancy is being considered. Follow-up genetic counseling may also be pursued for further evaluation and counseling of extended-family members. The nurse plays an instrumental role in reviewing the genetic summary with the patient and family and identifying information, education, and counseling needs for which follow-up genetic counseling might be useful.

Follow-Up at Various Life Stages

Prenatal Patients

A woman who has decided to continue a pregnancy with a fetus that has an open neural-tube defect requires follow-up genetic counseling so that coordination and collaboration of care throughout the pregnancy and delivery and in the newborn period are ensured. The genetics specialist arranges for the patient opportunities to meet or speak with other parents who have children with spina bifida, with the spina bifida clinic nurse specialist, with the medical and surgical specialists involved in ongoing care of the baby, and with other appropriate service providers. The arrangements are communicated to the primary care nurse and the team so that there is continuity of care and support.

Newborn Patients

For the family with a newborn with Down syndrome, follow-up genetic counseling might include a visit to the geneticist within the first few months after birth. The goal of this follow-up visit is to assess how the family is coping with the new diagnosis and to continue to develop and support an ongoing management plan in collaboration with the primary care provider.

Pediatric Patients

Children with cleft lip and palate require ongoing dental, hearing, speech, and oral surgical care. Genetic counseling is offered, often in conjunction with the comprehensive clinic visit, to follow up a diagnosis; the counselor seeks to answer family questions and to discuss with parents reproductive options and other genetic health concerns.

Adolescent Patients

Follow-up genetic counseling for the adolescent with Marfan syndrome is offered to review genetic information and to support integration of new genetic information into the patient's and family's daily life. During the follow-up visit, time is allowed for the adolescent to meet alone with the genetics specialist, so that the adolescent's questions and concerns can be addressed. The genetics specialist might discuss the importance of adhering to management and treatment recommendations and reinforce the option of identifying peers for support through the National Marfan Association. Concerns identified by the genetics specialist are conveyed to the primary care nurse and other team members (while respecting adolescent privacy rights), so that follow-up and support can continue.

Adult Patients

Follow-up genetic counseling for an adult with a family history of breast or other cancer may be necessary over a long period. Some patients use follow-up counseling sessions to continue discussion of their concerns and questions about presymptomatic testing, in preparation for making a decision regarding whether to pursue testing. Other patients request follow-up genetic counseling that includes extended-family members to discuss the inheritance factors and risks of cancer. Still others want to address reproductive concerns. The genetics specialist always offers to remain available for additional questions, concerns, or counseling needs and conveys this to the primary care nurse and provider.

Follow-Up Without a Specific Diagnosis

In some situations, a specific genetic diagnosis cannot be made, which may cause emotional distress for patients and families. Families may have long been seeking answers to such genetic health concerns as the cause of mental retardation in a family member. These families require special support, because they may not have specific genetic information with which to make reproductive or other health decisions. The genetics specialists review with these families the fact that a clinical diagnosis cannot be made in approximately one-half of individuals who have mental or physical disabilities but explain that new genetic technologies are expanding the possibilities for genetic diagnoses (Rimoin et al., 1996).

Follow-Up with New Treatment Options

Management of genetic disorders may involve new and complex treatments, including gene therapy. Patients diagnosed with cystic fibrosis, for example, may be offered participation in gene-therapy research trials. The genetics and medical teams following such patients communicate to the patients, nurses, and primary care providers involved the known short- and long-term risks, benefits, and limitations of these treatments. Many gene-therapy trials provide for only limited follow-up of such patients over time. The nurse may therefore have a central role in ensuring continuity of care, especially of children, as patients move through the health care system.

Follow-Up Resources for Patients and Providers

The nurse who provides ongoing care to patients and families with genetic conditions collaborates with the genetics team to identify and support ongoing genetic counseling needs. Genetics specialists are always available to both patients and primary care providers to address emergent genetic questions and concerns and to offer additional medical genetic information, management, and educational resources.

Nurses, as advocates, must be knowledgeable about the services and resources available in their communities and be able to recognize when gaps exist and to assist with referral. Coordination and collaboration with genetics specialists are essential nursing functions and ensure that the most comprehensive care is provided to families with current and future genetic health concerns. Nurses' participation in the genetic-counseling process is certain to expand as newer methods of diagnosis, treatment, and management are identified (Williams & Lea, 1995). Current aspects of nursing participation are summarized in Table 5.5.

TABLE 5.5

Nurses' Participation in Genetic Counseling

Assess family and psychosocial histories	Support nondirective counseling
Make referrals for genetic evaluation and counseling	Collaborate with genetics specialists for ongoing counseling needs
Provide appropriate information regarding the contribution of genetics to the health-illness continuum	Provide information regarding local and community support groups
	Participate in management of patients and families with identified genetic predispositions and conditions

CONCLUSION

Genetic counseling is a communication process that involves assessment, diagnosis, education, support, and follow-up. The goal of genetic counseling is to communicate to patients and their families information that will be most helpful as they make reproductive and health decisions and attempt to integrate new genetic information into their daily lives. Nurses in all areas of practice can expect to care for patients who have current or future genetic health concerns. Their participation in genetic counseling will be characterized by

- helping patients and families to recognize the contribution of genetics to health and disease;
- preparing patients and families for genetic counseling by explaining the purpose and goals of such counseling;
- offering support throughout the genetic-counseling process; and
- determining patients' ongoing genetic counseling needs and referring patients for follow-up as appropriate.

SUMMARY POINTS

Genetic counseling is expected to expand as the genetic contribution to the health-illness continuum is recognized.

Genetic counseling involves a communication process by which patients and families are helped to learn and understand relevant aspects of genetics, to identify options available for their decision making, and to receive support to incorporate personal and family genetic information into their daily lives.

Genetic counseling is available throughout the United States; services can be located through local, regional, and national professional genetics organizations.

The process of genetic counseling consists of five stages: assessment, analysis of genetic information, communication of genetic information, supportive counseling, and follow-up.

Nurses will be involved in the genetic-counseling process in carrying out the following activities:

- assessing and referring patients and families who may benefit from genetic counseling,
- offering anticipatory guidance by explaining the purpose and goals of genetic counseling,
- participating in family history and physical assessment,
- collaborating with genetics specialists to provide supportive and follow-up genetic counseling, and
- coordinating genetic health care and ensuring that patients have access to community and national support resources.

QUESTIONS FOR CRITICAL THINKING

1. Consider at least two patients in your practice who have (or may have) a genetic condition. List four ways that you might participate in genetic counseling in their care.

2. Ms. C comes to see you for her first prenatal visit. As you are taking her family history, she informs you that she and her partner are first cousins. She also notes a family history of hearing loss—

in herself, her brother, and several maternal aunts and uncles. Draw the family genetic pedigree and discuss actual and potential genetic-counseling issues. How would you explain these issues to Ms. C?

3. Nondirective genetic counseling is important to ensure patient autonomy and informed health decisions. Can you think of a patient counseling situation in which nondirective genetic counseling might *not* be appropriate? (Refer to Chapter 8 for further discussion.)

REFERENCES

Alliance of Genetic Support Groups (1995). *Directory of national genetic voluntary organizations and related resources.* Published by the Alliance of Genetic Support Groups, Chevy Chase, MD.

American College of Obstetrics and Gynecology. Committee on Obstetrics: Maternal and Fetal Medicine. (March, 1993). *Folic acid for the prevention of recurrent neural-tube defects.* (Comm. Opin. No. 120). The American College of Obstetricians and Gynecologists, 409 12th St. SW, Washington, DC.

Andrews, L. B., Fullarton, J. E., Holtzman, M. A., and Motulsky, A. G., eds. (1994). *Assessing genetic risks: implications for health and social policy.* Washington, DC: National Academy Press.

Bennett, R. L., Steinhaus, K. A., Uhrick, S. P., et al. (1995). Recommendations for standardized human pedigree nomenclature. *Am. J. Hum. Genet.* 56:745–752. Reprinted in the *J. Genet. Counsel.* 4(4):167–279.

Biesecker, B. B. (1997). Psychological issues in cancer genetics. *Semin. Oncol. Nurs.* 13(2): 129–134.

Children's PKU Network (CPN), 10515 Vista Sorrento Parkway, #204, San Diego, CA 92121. Telephone: 619-587-9421; fax: 619-450-5034.

Clubb, R. L. (1991). Chronic sorrow: adaptation patterns of parents with chronically ill children. *Pediatr. Nurs.* 17(5):461–465.

Cohen, F. (1984). *Clinical genetics in nursing practice.* Philadelphia: Lippincott.

Council of Regional Networks. (1997). Guidelines for clinical genetic services for the public's health. 1st ed. Atlanta, GA: Council of Regional Networks.

Epstein, C., et al. (1975). Genetic counseling (statement of the American Society of Human Genetics Ad Hoc Committee on Genetic Counseling). *Am. J. Hum. Genet.* 27:240–242.

Fine, B. A. (1993). The evolution of nondirectiveness in genetic counseling and implications of the Human Genome Project. In D. M. Bartels, B. S. LeRoy, and A. L. Caplan (eds.), *Prescribing our future: ethical challenges in genetic counseling,* pp. 101–118. New York: Aldine de Gruyter.

Fisher, N. (1995). *Multicultural awareness in genetic counseling: a guide for genetics professionals.* Baltimore: Johns Hopkins University Press.

Forsman, I. (1994). Evolution of the nursing role in genetics. *J. Obstet. Gynecol. Neonatal. Nurs.* 23(6):481–486.

Harper, P. S. (1993). *Practical genetic counseling.* (4th ed.). Oxford: Butterworth-Heinemann.

Harrison, M. R., Golbus, M. S. and Filly, R. A. (1991). *The unborn patient.* Philadelphia: Saunders.

Minnick, M. A., Delp, K. J., and Ciotti, M.C. (1996). *A time to decide, a time to heal: for parents making difficult decisions about babies they love* (4th ed.). St. John's, MI: Pineapple Press.

National Marfan Foundation (NMF), 382 Main Street, Port Washington, NY 11050. Telephone: 800-862-7326; fax: 516-883-8712.

National Neurofibromatosis Foundation, Inc. (NNFF), 95 Pine Street, 16th Floor, New York, NY 10005, 212-344-NNFF or 1-800-323-7938, fax 212-747-0004.

National Tay-Sachs and Allied Disease Association, Inc. (NTSAD), 2001 Beacon Street, Room 204, Brookline, MA 02146. Telephone: 617-277-4463; fax: 617-277-0134.

Peters, J. A., and Stopfer, J. (1996). The genetic counselor's role in familial cancer. *Oncology* 10(2):159–166.

Reed, S. L. (1980). *Counseling in medical genetics.* 3rd ed. New York: Alan Ruliss.

Rimoin, D. L., Connor, J. M., and Pyeritz, R. E. (1996). Emery-Rimoin's principles and practice of medical genetics (3rd ed.). New York: Churchill Livingstone.

Sorenson, J. R. (1993). Genetic counseling: values that have mattered. In D. M. Bartels, B. S. LeRoy, and A. L. Caplan (eds.), *Prescribing our future: ethical challenges in genetic counseling,* pp. 3–14. New York: Aldine de Gruyter.

Spina Bifida Association of America (SBAA), 4590 MacArthur Boulevard, NW #250, Washington, DC 20007-4226. Phone: 800-621-3145; fax: 202-944-3295.

Williams, J. K., and Lea, D. H.(1995). Applying new genetic technologies: assessment and ethical considerations. *Nurse Pract.* 20(7):16–46.

CASE STUDY 5.1

PRECONCEPTION COUNSELING: ANTICIPATORY GUIDANCE

Mr. and Mrs. T come to the prenatal clinic to request a preconception visit with their nurse-midwife. Mrs. T reports to the nurse that she is concerned because of a family history of Tay-Sachs disease;* her sister died during infancy from complications of the disorder. Mrs. T explains that both she and Mr. T are of Eastern European Jewish descent. Mrs. T says that she is aware that people from this ethnic background have a higher chance of being carriers of a gene for Tay-Sachs disease. Mr. and Mrs. T ask the nurse about the probability of their being carriers of Tay-Sachs disease, their likelihood of having a baby with Tay-Sachs disease, and whether testing is available to them during a pregnancy.

Tay-Sachs disease is an inherited condition that occurs more commonly among North American people of Ashkenazi (Eastern European) Jewish descent. The clinical features of Tay-Sachs disease include severe mental retardation and physical deterioration beginning in infancy; death occurs at age 2–3 years. Tay-Sachs disease occurs in approximately 1 in every 3,600 live births to people of Ashkenazi Jewish descent and in about 1 in 360,000 people in most other populations.

Tay-Sachs disease is inherited in an autosomal-recessive manner. The basic gene defect causes deficiency or absence of the enzyme hexosaminidase A, which leads to an accumulation of fatty substances along the neural pathways. Treatment is symptomatic, and there is no cure. Carrier and prenatal testing are available for Tay-Sachs disease (Harper, 1993; Rimoin et al., 1996).

- What information might you provide to help answer Mr. and Mrs. T.'s questions?
- How would you describe genetic counseling?
- What additional information might you gather in preparation for genetic counseling and testing?

Discussion

The nurse could explain that because Mrs. T.'s sister died from Tay-Sachs disease, Mrs. T.'s chances of being a carrier of the gene for Tay-Sachs disease are increased. The nurse might note that carrier and prenatal testing are available for this disease and that further information about the testing will be provided by the physician when he or she meets with the couple. Patients who have genetic questions or concerns are often referred to a regional genetics clinic for additional information and counseling. The nurse should discuss genetic counseling, explaining that it is a communication process designed to provide information to patients to help them in making health and reproductive decisions. The genetics professional, the nurse explains, will provide information about Mr. and Mrs. T.'s chances of having a baby who is affected with Tay-Sachs disease and about the risks, benefits, and limitations of carrier and prenatal testing. The nurse can inform Mr. and Mrs. T that a complete family history will be taken at the time of the genetic-counseling visit and that, whenever possible, the genetics specialist will attempt to obtain for complete evaluation any medical records regarding the genetic condition in the family. The nurse should also inform the couple that they can get further information about Tay-Sachs disease from the National Tay-Sachs and Allied Disease Association (National Tay-Sachs and Allied Disease Association).

PRENATAL DIAGNOSIS: PROVIDING SUPPORTIVE COUNSELING

Mrs. P is a pregnant 22-year-old patient. During a routine ultrasound examination performed in her physician's office at 18 weeks' gestation, a large, lumbar neural-tube defect* is observed in the fetus. The physician refers Mrs. P and her husband to

Neural-tube defects occur when the neural tube, which forms the spinal column and spinal cord, does not close completely during the very early weeks of pregnancy. Such defects, including anencephaly and spina bifida, are common and occur in approximately 1 in every 1,000 live births. Neural-tube defects can be detected by prenatal testing, using such methods as maternal serum alpha-fetoprotein analysis, ultrasound examination, and amniocentesis (Harrison et al., 1991).

a perinatal center for further evaluation and counseling. Mrs. P, understandably distraught, asks the nurse what she can expect at the perinatal center and expresses a desire to have her parents attend the session. Mrs. P tells the nurse that she is afraid to undergo amniocentesis because "it might give me even worse news." She asks the nurse, "What would you do if you were in my situation, and what are our options?"

- What other questions can you anticipate that Mrs. P and her husband might ask?
- What information might you provide to this couple about the amniocentesis procedure?
- What additional support could you offer to Mrs. P and her husband?
- How might you respond to Mrs. P's questions about what she should do and what her options are?

Discussion

The nurse can anticipate that this couple will experience shock and may express denial, anger, and guilt as they incorporate this unexpected information into their lives. Questions often asked by couples who receive a prenatal diagnosis of a fetal abnormality are Why did this happen? Did we do anything to cause this to happen? Is there any treatment available? What are the choices available to us at this time? Will this happen again?

The nurse can provide support by acknowledging Mr. and Mrs. P's shock and concern, noting that these reactions are a normal part of receiving unexpected news. She or he can provide guidance that anticipates other reactions by explaining to Mrs. P and her husband that they may experience feelings of grief on having to abandon certain expectations for their baby. The nurse should also explain that a genetic counselor, genetic nurse specialist, and geneticist will be available at the perinatal center to help answer their questions and that Mrs. P's and her husband's parents are welcome to be a part of the genetic consultation. The nurse, being aware that it is critical for the couple, on the basis of personal beliefs and values, to decide on a course of action regarding the pregnancy, explains that it is not her (or his) practice to tell patients what she (or he) would do in their situation but that she (or he) *can* offer as much information and support as possible to assist the couple in understanding their baby's condition and in making informed choices.

The nurse can also explain that high-resolution ultrasound examination at the perinatal center will be helpful in further evaluating the nature and extent of the neural-tube defect. The nurse could explain that amniocentesis is another available prenatal test, which requires insertion of a needle under ultrasound guidance to obtain fluid for analysis and evaluation of the defect. The patient should be informed that genetic specialists and a perinatologist will provide additional details about the risks, benefits, and limitations of this testing.

The nurse, as a final statement, should tell the couple that additional information is available from other specialists such as neurologists and from local and national support organizations such as the Spina Bifida Association of America. Parents who have been through a similar experience might also be contacted for support (Spina Bifida Association of America).

The nurse should note in the chart that she or he will discuss with the family the role that folic-acid supplementation, prescribed prior to conception and taken during the first 3 months of any subsequent pregnancy, plays in significantly reducing the risk of neural-tube defects (ACOG, 1993).

CASE STUDY 5.3

NEWBORN SCREENING: MAKING A GENETIC REFERRAL

Baby girl R, screened at birth for phenylketonuria (PKU)*, is diagnosed with the condition. Mr. and Mrs. R bring the baby to the pediatrician for her 6-week health

Phenylketonuria is an inherited metabolic condition that occurs in approximately 1 in 10,000 live-born infants. PKU is inherited in an autosomal-recessive manner and is caused by a gene defect that interferes with the body's ability to break down phenylalanine to tyrosine. Because it cannot be degraded by people who have PKU, phenylalanine accumulates in body fluids and damages the developing central nervous system in early childhood. Untreated, it causes severe mental retardation. The treatment of PKU consists of dietary modification to prevent phenylalanine accumulation. Newborn screening for PKU is the accepted practice (Harper, 1993; Rimoin et al., 1996).

examination. At the visit, the nurse notes that Mr. and Mrs. R recently moved to the United States from Poland. Mrs. R does not speak English. Mr. R tells the nurse that he and his wife do not understand what the diagnosis of PKU means for their baby. He asks, "Will she die?" He also tells the nurse that his wife believes that God is punishing them and that this is why their baby requires a special diet. He asks whether the nurse can help to address his concerns.

- How might you explain to Mr. R the inheritance of PKU and the meaning of this diagnosis?
- How would a referral for additional genetic counseling be helpful to this family?
- What other information would you gather about the family in preparation for genetic counseling?

Discussion

The nurse can offer education and support in this counseling setting. The nurse could explain that PKU is an inherited genetic condition, in which each parent is a carrier of a single altered gene. Knowing the diagnosis of PKU may be upsetting, but this knowledge also is helpful, because treatment is available that can prevent the serious consequences of this condition. The nurse should reinforce that Mr. and Mrs. R's baby is receiving the correct treatment and confirm the importance of continuing to follow the prescribed diet. The nurse, being aware of potential cultural influences on dietary practices, can offer to set up an appointment with a dietician who is familiar with these issues and who can help the family to tailor baby R's diet to meet these needs.

The nurse relates the parents' concerns to the pediatrician, and together they decide to offer genetic counseling to Mr. and Mrs. R for further support and explanation. The nurse could involve in the counseling session a priest, who might talk with Mrs. R about her concerns that she is being punished by God. The nurse can tell Mr. and Mrs. R that genetic counseling will provide them with additional information about PKU, including how the disease is inherited, and that it will offer the Rs an opportunity to explore their concerns about cause. The nurse should assure Mr. R that an interpreter will be available to them, so that Mrs. R can understand what is being said. In addition, the nurse might provide Mr. and Mrs. R with information about the Association for Neuro-Metabolic Disorders, which offers parent information, peer support, and matching of individuals and families, and a telephone help line; these services are often available in other languages (Children's PKU Network).

CASE STUDY 5.4

GENETIC DIAGNOSIS: PREPARING THE PATIENT FOR GENETIC EVALUATION AND COUNSELING

A 14-year-old girl, Susan, is brought to the clinic by her mother for a physical examination in preparation for sports camp. The nurse notes during examination that Susan's height is greater than the ninety-fifth per- centile and that she has long, thin fingers, scoliosis, a concave chest, and a heart murmur. The mother volunteers that Susan's father is 6 feet 5 inches tall and also has a heart murmur and that the paternal grandmother died suddenly in her thirties (40 years ago). The nurse enters these findings in the chart and discusses them with the physician. The physician suspects that

Susan may have Marfan syndrome* and recommends a genetic evaluation and counseling. The physician asks the nurse to provide the family with details about the genetic consultation. Susan's mother has many questions about Marfan syndrome.

- What information about Marfan syndrome might you discuss with Susan and her mother?
- What could you tell Susan and her mother to expect from genetic evaluation and counseling?
- What additional information would you gather for the genetic counseling consultation?

Marfan syndrome is a disorder of fibrous connective tissue with characteristic findings of three body systems: skeletal, ocular, and cardiovascular. People with Marfan syndrome are usually taller than the ninety-seventh percentile for their age and have long limbs and extremities, long fingers, a long face, chest concavity (pectus excavatum), and dislocated lenses. Manifestations among affected persons vary, with some patients experiencing severe clinical problems such as aortic root dissection, which can lead to sudden death. No specific therapy exists for the underlying gene defect in Marfan syndrome. Treatment efforts are directed at determining which problems are present, anticipating the problems that will probably arise in the future, and pursuing prophylactic measures for specific problems (Rimoin et al., 1996).

Discussion

The nurse should explain that the term *syndrome* refers to a collection of characteristics that occur in a person and are believed to have a single or common cause. Marfan syndrome may be inherited in some families, as an autosomal-dominant condition, or it may occur in a single individual. The diagnosis of Marfan syndrome is currently based on the presence of clinical findings. Genetic evaluation and counseling is carried out by specially trained professionals, including medical geneticists who have specialized in the diagnosis of genetic syndromes. The nurse could relate that genetic evaluation and counseling for Marfan syndrome involves a review of family history, medical records, and physical examination. Additional evaluations include an echocardiogram for further assessment of the heart and aortic root, and an eye examination to check for lens dislocation. Because Marfan syndrome may be inherited in some families, evaluation may be offered to other family members. The nurse can also provide the family with the address and telephone number of the National Marfan Foundation, which can provide additional information about the condition (National Marfan Foundation).

CASE STUDY 5.5

FOLLOW-UP AFTER A GENETIC DIAGNOSIS

A 3-year-old boy with mild developmental delays is seen in the pediatric clinic for an annual checkup. In reviewing his records prior to this visit, the nurse notes that the boy has recently received a genetic diagnosis of neurofibromatosis type 1 (NF-1)*. The medical summary sent to the clinic by the geneticist and genetic counselor states that the diagnosis of NF-1 was made on the basis of clinical findings: seven café-au-lait spots, axillary freckling, Lisch nodules seen on ophthalmological examination, and several neurofibromas. Magnetic resonance imaging of the boy's brain did not reveal any evidence of tumors. The geneticist recommended that the boy's parents and two sisters undergo a genetic evaluation to rule out NF-1 in each of them. At the time of her examination, the

Neurofibromatosis type 1, also known as von Recklinghausen's disease, is a common disorder of the nervous system. In many families, NF-1 is inherited in an autosomal-dominant manner, but approximately half of the cases of NF-1 result from a new gene mutation. People who have NF-1 demonstrate a wide variety of symptoms, or variability of expression. Common features include café-au-lait spots, fibromatous skin tumors, typical ocular findings in the iris known as Lisch nodules, and an increased risk of malignant tumors. Management of patients with NF-1 involves annual clinical evaluations to detect complications beginning in the first 2 years of life (Rimoin et al., 1996).

boy's mother asks the nurse to explain why she and her family must return for further genetic evaluation. She says that this new diagnosis is enough to cope with and that she does not want anyone else in the family to know of the history.

- What could you tell the mother about NF-1 and its inheritance in families?
- How could you explain the recommendation for further evaluation of other family members?
- What additional nursing interventions would be helpful to support family members as they integrate this new information?

Discussion

The nurse could explain that NF-1 is often inherited in families and that, when a diagnosis of NF-1 is made in a person, it is standard medical genetics practice to recommend that other family members be offered a genetic evaluation to determine whether they too are affected. Further evaluation in this family will help to identify whether NF-1 is a hereditary condition in the family or whether it occurred as a result of a new gene mutation. This knowledge will be helpful in further evaluating the patient's siblings. Appropriate medical management can be offered to others in the family who are found to have NF-1. Further assessment of the parents may also be helpful in reproductive and family health planning.

The nurse can support the mother by explaining that learning of a new genetic diagnosis in a family member is often difficult for a family, especially if the news is unexpected. Family members may need time to incorporate new genetic information into their daily lives. Finally, the nurse can offer information and support to the family through the National Neurofibromatosis Foundation (Alliance of Genetic Support Groups; National NF Foundation).

The nurse should explain to the mother that all information regarding her son's diagnosis of NF-1 is confidential and cannot be shared with other people, including family members, without her permission. The nurse should also explain, however, that it is important for the mother to consider discussing the diagnosis with other family members, so that they can have an equal opportunity to learn more about this condition and the potential for them and their children to develop it.

SUPPORTIVE CARE AND MANAGEMENT

The nurse is caring for 16-year-old Martha, who has a history of mild mental retardation.* Although the family has been to several genetics diagnostic clinics and Martha has undergone many genetic tests, a specific cause of her mental retardation has not been discovered. Martha's mother

*The prevalence of mental retardation is approximately 3 percent in the United States, and more males than females are affected. Approximately 1 in 10 Americans has a relative who is mentally retarded. Causes of severe mental retardation are more well defined than are causes of mild mental retardation, which may result from multiple genes and other familial factors. In approximately 60 percent of people with mild mental retardation, a specific diagnosis is not identified (Harper, 1993; Rimoin et al., 1996).

informs the nurse that Martha's 18-year-old sister is currently in a serious relationship with a young man, and the mother wonders about the sister's chance of bearing children with a similar condition. Martha's mother tells the nurse that she is considering suggesting to her 18-year-old daughter that she never have children so that "she doesn't go through all of the heartache and guilt I did." The mother asks the nurse where she can go for further information.

- How might you respond to Martha's mother?

- How might genetic counseling be helpful to this family, and how would you explain this to Martha's mother?

- What other support could you offer to the family?

Discussion

The nurse might respond to Martha's mother by acknowledging that not understanding the basis of Martha's mental retardation can be difficult and that it is not unusual for parents to experience guilt when they have a child with a disability. The mother's feelings of guilt could be explored to try to achieve an understanding of the cause of these feelings. The nurse can reinforce, however, that often a specific diagnosis for mild mental retardation cannot be discovered, though newer genetic technologies are helping to expand our understanding of the causes of mental retardation.

The nurse can offer genetic counseling to the family to evaluate the sister's chances for bearing children who are mentally retarded. She or he should explain that genetic counseling involves analysis of family history and medical histories and that it might provide additional information that will help Martha's sister come to her own decision about reproduction. Furthermore, review of the family and medical history with Martha's mother may be helpful to her in uncovering the reasons for, and dispelling, her feelings of guilt. The nurse might explain that the regional genetics clinic has a family support group for family members of people with unexplained mental retardation and other conditions.

The Process of Genetic Testing for Nurses and Their Patients

The wider paradigm of relationships and family transcends the old group definitions. The discovery of our connection to all other men, women, and children joins us to another family. Indeed, seeing ourselves as a planetary family struggling to solve its problems, rather than as assorted people and nations assessing blame or exporting solutions, could be the ultimate shift in perspective.　　　　*Marilyn Ferguson*

GENE CARE LINK

For a comprehensive view of genetic-testing issues, go to http://www. jbpub.com/ clinical-genetics.

OBJECTIVES
- Describe key issues for patients in the genetic-testing process.
- Discuss the role of the nurse in offering, reporting, and interpreting genetic tests.

RATIONALE
Nurses will be called upon to assist with offering, reporting, and interpreting genetic tests to individuals and families and will therefore need to understand the genetic-testing process and its effects on patients' perceptions of health and risk, family dynamics, and coping mechanisms.

APPLICATION ACTIVITIES
- Identify individuals and families to whom molecular genetic testing should be offered.
- Participate in pretest assessment, counseling, and follow-up during the genetic-testing process.
- Collaborate with other health professionals in providing care to individuals and families who are undergoing genetic testing.

INTRODUCTION

G*enetic testing involves the use of specific laboratory assays to determine whether a patient who, because of family history or clinical symptoms, has been identified as being at higher risk for a particular inherited trait or condition may have a gene for that condition (Andrews et al., 1994). Genetic testing using molecular techniques is accomplished either by direct mutational analysis or by indirect methods, each having different levels of sensitivity and specificity. As genetic testing becomes more widely applied, provision of appropriate and current information to patients deciding whether to undergo testing, accurate interpretation of test results, and support of patients and families throughout the testing process will be essential, because people often base critical health decisions on this information.*

This chapter describes genetic-testing approaches that use molecular techniques and some of the more common clinical applications of such tests. We will address the use of genetic testing, nursing roles, and patient issues in the areas of prenatal diagnosis and reproductive options, carrier testing, genetic diagnosis when preventive medical interventions are available, and predictive testing for genetic conditions for which there is currently no specific treatment or cure. We will present case studies to illustrate the issues considered in the use of genetic tests in these clinical situations.

GENETIC-TESTING METHODS

The scope of laboratory techniques to detect genetic disorders has greatly expanded during recent years. As a result, nurses are participating in offering and interpreting different genetic tests to patients and families. Familiarity with the genetic-testing process will facilitate integration of this new knowledge into nursing practice.

Molecular genetic testing is currently available for carrier testing, prenatal diagnosis, confirmation of a genetic diagnosis, and for genetic susceptibility or predisposition testing. Carrier testing using DNA methodologies is performed to identify apparently healthy individuals who carry a single copy of an altered gene that, when present in two copies, causes a recessively inherited genetic condition: cystic fibrosis carrier testing falls into this category. A molecular genetic test may be performed to identify a genetic predisposition to a late-onset disorder for which preventive measures are available: testing individuals in families with familial adenomatous polyposis is an example of this use. Predictive testing for late-onset genetic conditions such as Huntington's disease is now available. Although for this and other late-onset conditions, no specific treatment or cure currently exists, knowledge about the presence of a gene that predisposes an individual to develop Huntington's disease can be helpful for reproductive and life-style planning. Additional uses of genetic testing include monitoring previously diagnosed conditions such as familial cancers. Information gained from these genetic tests may help in identifying those patients who are at higher risk for recurrence and those who need more intensive therapies. Two general approaches—direct and indirect genetic tests—are currently used to identify gene mutations and predispositions.

Direct Genetic Testing

Direct molecular genetic testing can be done in situations in which the precise gene mutation has been identified. Once a precise gene mutation is known, patients who have a family history of the genetic condition may be tested directly for the presence or absence of the specific gene mutation. Genetic testing for the gene mutation in these situations has a high degree of accuracy when performed by an experienced molecular laboratory. When a direct genetic test is used, a blood or tissue sample is obtained and analyzed only from the individual who has an increased risk for the condition. Direct tests are now being used for cystic fibrosis, fragile X syndrome, and Huntington's disease, as well as for many other single-gene conditions (Andrews et al., 1994; Gusella & MacDonald, 1994; Schulman et al., 1996; Tarleton & Saul, 1993; Warren & Nelson, 1994; Table 6.1).

TABLE 6.1

Examples of Genetic Conditions for Which Direct DNA Testing Is Available

Achondroplasia	Huntington's disease
Breast cancer	Myotonic dystrophy
Colon cancer	Sickle-cell anemia
Cystic fibrosis	Tay-Sachs disease
Duchenne muscular dystrophy	Thalassemia
Fragile X syndrome	

Source: Helix, National Directory of Diagnostic Laboratories, Children's Hospital and Medical Center, P.O. Box 5371, CH-94, Sand Point Way, NE, Seattle, WA 98105-0371.

Indirect Genetic Testing

Indirect methods of molecular genetic testing are less precise and require more complex interpretive analyses than do direct methods. This is because indirect molecular genetic testing relies on identifying whether a person has inherited a region of a specific chromosome that contains a gene rather than on identifying the gene mutation itself. In many instances, prior to the discovery of the specific gene mutation that causes a genetic condition, genetic testing involves localization of the gene to a particular chromosome and then to a particular segment of the chromosome. This process, accomplished by linkage analysis, allows for the indirect detection of a gene.

During meiosis, maternal and paternal chromosomes normally line up next to one another, and an exchange between portions or segments of the paired chromosomes occurs. This normal process is called *recombination*. Recombination accounts for the unlimited variation observed in humans, such that each person's genetic material is unique. The process of recombination is also the basis for understanding linkage analysis. Known sequences of DNA near the gene of interest are used as biological markers that can track the gene in a family from generation to generation. The closer the known gene marker is to the disease gene, the greater the likelihood that the person inherited the marker and disease gene together. The farther apart the gene marker is from the disease gene, the greater the possibility that a recombination event occurred, which could lead to incorrect interpretation of the test results. This possibility is included in the interpretation of the genetic test results as a probability that a member of a family may have inherited the disease gene. Results of linkage studies may therefore be a less definite determination of a person's gene status than is direct mutational analysis.

Genetic-linkage studies require the participation of family members across generations in addition to the family member who has an increased chance of inheriting the gene in question. Usually, testing of one or more affected and unaffected family members is required to determine whether an individual family member has inherited the gene.

Although useful when direct testing is not possible, linkage analysis has several limitations. Usually, DNA must be available from at least one affected family member. This may be obtained from either a living family member or from a stored tissue sample. In every case, permission from the affected individual or guardian is needed before testing can be performed. Other unaffected family members may also have to undergo testing; again, permission from these family members is necessary. Maternity and paternity must be known to allow proper interpretation of results. Another limitation of linkage analysis is the possibility of inconclusive information resulting from recombination, as described earlier. Markers may be uninformative, again leading to inconclusive results.

In the past, linkage analysis was used for conditions such as Huntington's disease before direct DNA analysis became available (Jackson, 1987). It is still being used for conditions such as Marfan syndrome, neurofibromatosis, and other genetic conditions for which multiple mutations are known, most of which are unique to individual families (Table 6.2).

TABLE 6.2

Genetic Conditions for Which Linkage Analysis Is Used

Familial melanoma

Hemophilia

Marfan syndrome

Neurofibromatosis

Polycystic kidney disease—adult onset

Source: Helix, National Directory of Diagnostic Laboratories, Children's Hospital and Medical Center, P.O. Box 5371, CH-94, Sand Point Way, NE, Seattle, WA 98105-0371.

TABLE 6.3

Nursing Roles in Offering and Interpreting Genetic Testing

Identify patients who, because of family or medical history, may benefit from genetic testing.

Refer patients and families for further evaluation and counseling as needed.

Collaborate with other health professionals, including genetic specialists and testing centers, to provide information and support.

Participate in discussion with patients regarding benefits and limitations of genetic testing.

Assure informed consent.

Advocate for patients' rights to choose or to refuse to undergo genetic testing.

Provide ongoing evaluation of patients' understanding of and coping with the testing process.

Source: Abstracted from Scanlon & Fibison, 1995.

Complexities of Genetic Testing

Genetic testing is extremely complex. Knowledge and understanding of the different methods used for genetic testing and of their applications will help nurses to provide accurate, complete information to patients and families and to support informed health decisions. Patients and families often use genetic testing to make decisions regarding reproductive choices such as prenatal diagnosis, whether to have predictive testing when a treatment or cure is not available, testing of children, and choice of treatments, including gene therapy. Everyone on a genetics health-care team must be aware of the benefits and limitations of genetic testing and the potential impact of new genetic information on patients and families. Examples of nursing participation in genetic testing and interventions are outlined in Table 6.3.

The remainder of this chapter provides examples of genetic conditions and discussion that illustrate some of the psychological, social, and ethical issues that patients and nurses may face during the genetic-testing process. It also describes the role of the nurse in a variety of settings and in collaboration with other health professionals. Case studies illustrating nursing participation in the testing process are also provided.

GENETIC-TESTING ISSUES

Genetic Testing for Confirmation of Diagnosis, Carrier Status, and Prenatal Diagnosis: Fragile X Syndrome

Fragile X syndrome is currently recognized as one of the most common forms of inherited cognitive disability, occurring in approximately 1 in 1,250 males and 1 in 2,000 females. In addition to mental retardation, patients with this condition often have other behavioral problems, such as hyperactivity, disordered speech, and autistic behavior. Although variable, typical physical features include a long narrow face, protruding ears, and abnormally large testes, which usually are not apparent until puberty (Hagerman & Silverman, 1991; Lessick, 1993; Nelson, 1993; Oostra et al., 1993).

Fragile X syndrome is caused by a gene mutation on the X chromosome. The major defect involves elongation of a repeated DNA sequence that is normally very short. This condition is referred to as a *full mutation*. New knowledge of the gene defect present in fragile X syndrome has helped to clarify the atypical, complex, X-linked inheritance pattern and the varying degrees of severity in affected families (Rousseau, 1994).

Males and females may also be carriers of a moderately elongated DNA sequence that is caused by changes in the DNA without a gene alteration. This condition is termed *premutation*. Individuals who carry the premutation do not have the typical clinical features of the fragile X syndrome. Males who carry the premutation are called *transmitting males*. All of their daughters will inherit the premutation and are usually clinically normal. However, these daughters have an increased chance of giving birth to babies with disabilities, which may include mental retardation. With each pregnancy, a female who carries a premutation has a 50 percent chance (1 chance in 2) of passing on the fragile X gene alteration to her child.

Diagnostic and Reproductive Testing for Fragile X Syndrome: Methodology

In the past, laboratory diagnosis of fragile X syndrome relied on a special chromosome analysis for detection of a "fragile site" on the X chromosome. Currently, direct DNA testing is available for diagnosis of affected individuals, for carrier testing of unaffected family members, and for prenatal diagnosis in families with a positive family history. DNA testing for fragile X syndrome is more precise than is chromosomal analysis, especially for carrier detection. However, approximately 1 percent of cases will not be detected by DNA analysis. In such situations, analysis of samples from other family members may be helpful.

Prenatal diagnosis for fragile X syndrome usually involves analysis of fetal cells obtained by amniocentesis or chorionic villus sampling. Although helpful in prenatal detection of males with fragile X syndrome, prenatal diagnostic testing currently does not provide information about the clinical severity in approximately 50 percent of female fetuses identified as carrying the full mutation.

Current Recommendations for Diagnostic and Carrier Testing for Fragile X Syndrome

Guidelines for offering fragile X testing, developed by the American College of Medical Genetics (American Society of Human Genetics Board of Directors and the American College of Medical Genetics Board of Directors [ASHG/ACMG], 1995) currently state that DNA testing for fragile X syndrome be considered for

- any person who has unexplained mental retardation, developmental disabilities, or autism, especially if he or she has any of the physical and behavioral characteristics of the fragile X syndrome, a family history of fragile X syndrome, or relatives with undiagnosed mental retardation;
- individuals seeking reproductive counseling who have a family history of fragile X syndrome or a family history of unspecified mental retardation;
- any individual who has a clinical indication (including a risk for being a carrier) despite a negative or ambiguous fragile X chromosome study or who has had a previously positive fragile X chromosome analysis but who does not have the typical appearance for fragile X syndrome. (See Case Study 6.1.)

Presymptomatic Genetic Testing for Conditions for Which There Is No Effective Treatment or Cure: Huntington's Disease

In the years between locating the Huntington's disease (HD) gene on chromosome 4 and specifically identifying the gene, linkage analysis was used to identify patients who carried the gene for HD. The identification of a specific gene mutation for HD now permits direct testing. However, efforts to discover a cure for HD have not yet been successful. Therefore, direct genetic testing, even without involvement of other family members, is challenging with regard to the many complex ethical, legal, and social issues that must be discussed throughout the testing process (Andrews et al., 1994; Huntington's Disease Society of America, 1994; Wilke, 1995).

HD is an inherited, progressive, degenerative brain disorder, usually of late adult onset, that occurs in 1 in 10,000 people. It is estimated that some 30,000 people in the United States have symptoms of HD and an additional 150,000 have an increased chance for inheriting HD from a parent who carries the HD gene (Huntington's Disease Society of America, 1994; Rimoin et al., 1996).

HD is an autosomal-dominant disorder caused by a mutation in a single gene. Affected individuals usually experience involuntary movements of all body parts. They may also experience cognitive impairment and psychiatric disturbances that may be manifested as severe suicidal depression, apathy, or obsessive-compulsive behaviors. The onset of symptoms usually occurs between the ages of 30 and 50. However, symptoms may appear at any age. Early symptoms may vary and often go undetected. These include minor twitching, clumsiness, changes in judgment and memory, alterations in gait, and in some individuals, behavioral changes such as depression. The duration of symptoms ranges from 10 to 20 years, with progression to death, which is usually caused by heart failure or aspiration pneumonia. Currently, no effective treatment or cure exists (Hunt & Walter, 1991; Kovach & Stearns, 1993; Rimoin et al., 1996).

In 1983, scientists located the gene for HD on chromosome number 4. In 1986, DNA markers linked to the HD gene were discovered, making presymptomatic and prenatal testing available in families with a positive family history (Gusella & MacDonald, 1994). In 1993, 10 years after the discovery of its location, the HD gene was identified and isolated by the Huntington's Disease Collaborative Group. The gene alteration was found to be a small, expanded, and repeated DNA sequence referred to as a *CAG repeat* (Kremer et al., 1994). The highest number of the repeated DNA sequence has been reported in children; however, a correlation between numbers of repeats and the age of onset has not yet been completely determined (Andrews et al., 1994; Noremolle et al., 1995).

Presymptomatic Genetic Testing for Huntington's Disease: Psychosocial Considerations

Patients with a family history of HD have identified a number of reasons for choosing presymptomatic genetic testing. Some who choose to pursue testing do so in order to be able to make informed plans regarding such issues as marriage, reproduction, career, and finances. Others wish to be relieved of the uncertainty of being at risk. For these individuals, knowledge of their status, regardless of the test outcome, is more beneficial than ignorance of it (Huntington's Disease Society of America, 1994; Wexler, 1990). Many people have indicated a desire to know whether they carry the HD gene but, when presented with the opportunity to be tested, elect not to follow through, because the emotional experience or concerns about confidentiality outweigh the benefits of knowing their gene status.

A variety of responses have been observed in those who have chosen testing. Some have found the information helpful in determining the direction of their personal and professional lives, even when they have been found to be gene carriers. Others have experienced significant emotional trauma when they learn they do *not* carry the gene, because each has developed a personal identity based on the knowledge that he or she would be affected. Some people have been hospitalized for depression after learning that they carry the gene for HD, whereas others have described feeling more calm after resolving uncertainty (Hunt & Walter, 1991; Huntington's Disease Society of America, 1994).

A person's decision to pursue presymptomatic testing for HD should always be informed and freely made. The nurse can assure patients involved in the decision-making process that they have the freedom to choose and need not feel coerced into testing by a partner, another family member, a health care provider, insurance provider, or employer (Huntington's Disease Society of America, 1994).

Current Guidelines for Presymptomatic Testing for Huntington's Disease

The Huntington's Disease Society of America (1994), together with the World Federation of Neurology Research Group on Huntington's Disease and the International Huntington's Disease Association, have developed guidelines for presymptomatic testing for HD. The guidelines include indirect and direct testing for HD (*J. Med. Genet.*, 1990 & 1994; *J. Neurol. Sci.*, 1989). A team approach to presymptomatic testing for HD in a designated HD testing center is currently recommended. Primary care providers are advised to refer individuals interested in testing to the nearest designated center. HD testing centers offer trained personnel, who are best equipped to provide counseling and administer the recommended tests.

It is recommended that the following components be included in an HD testing program:

- an initial telephone contact that includes a prescreening interview with the person who is concerned about his or her risk for HD;
- three in-person, pretest sessions, during which are offered genetic counseling, neurological, and psychological examinations, and provision of additional information, such as reading materials, to further ensure informed decision making;
- a fourth, follow-up session to discuss test results;
- follow-up posttest counseling sessions over a 2-year period.

Prior to participation in the testing process, individuals are encouraged to identify a close friend or partner who can serve as a companion throughout the process. Identification of a counselor (psychologist, social worker, psychiatrist, or other mental health professional) nearby is also recommended, especially if the individual lives at a distance from the testing center.

Pretest counseling is considered the most important aspect of testing. Each person considering testing needs to be informed of

- the clinical, psychological, genetic, and reproductive implications of HD and available options;
- their current chance of inheriting the HD gene;
- the options for testing;
- the limitations of current testing, especially the possibility of ambiguous test results, and the accuracy of test results;
- the potential negative consequences of testing;
- the need for careful consideration of the risks, benefits, and limitations of testing, especially the implications of testing outcomes, either positive or negative, for the future.

Testing of minors for HD is not currently recommended, unless a medically compelling reason, such as the appearance of symptoms, is recognized. Such testing under all other circumstances is not usually considered to be in the child's best interest. However, differences of opinion regarding testing of children for HD remain, and testing centers are currently advised to develop their own policies with regard to the testing of minors (Huntington's Disease Association of America, 1994; Wertz et al., 1994). (See Case Study 6.2.)

Presymptomatic Genetic Testing for a Condition for Which Preventive Measures Are Available: Familial Adenomatous Polyposis

Genetic testing is now available to provide information that may clarify, prior to the onset of diagnostic symptoms, whether a person has inherited a gene for a condition. Such testing may be helpful to patients in making health decisions about inherited conditions for which prevention and treatment measures are available. Testing for the presence of the gene for familial adenomatous polyposis (FAP), an inherited form of colorectal cancer, is one example of this use of genetic testing.

Colorectal cancer is the second most common type of cancer in the United States today. Although in most families colorectal cancer occurs as a sporadic condition with no clear inherited predisposition, family studies have demonstrated that first-degree relatives (parent, child, sibling) of affected individuals have a twofold-to-fourfold increased chance for developing colorectal cancer (Bishop & Thomas, 1990). Less frequently, colorectal cancers may be inherited in an autosomal-dominant manner in families. Although they account for a smaller percentage of total cases of colorectal cancer, these families represent an important group that must be offered appropriate assessment, testing, and surveillance for early intervention and treatment.

FAP is one of the inherited colorectal cancers for which presymptomatic genetic testing is now available. FAP is an autosomal-dominant disorder occurring in 1 in 7,000 people and is caused by a single gene. An individual who inherits the FAP gene is predisposed to develop large numbers of adenomatous polyps throughout the colon which, if not treated, may lead to the development of colorectal cancer at an early age. The FAP gene is highly penetrant, meaning that 100 percent of individuals who have the gene will develop colorectal cancer by age 50 (Bulow et al., 1990; Schneider, 1994). In addition to being at increased risk for developing cancer, individuals with FAP may also develop osteomas and subcutaneous cysts. The average age at diagnosis of cancer in people with FAP is approximately 36 years; however, polyps may develop within the first decade of life. Therefore, early detection and treatment of those individuals who have inherited the FAP gene provides an important means of preventing excessive mortality from this form of colorectal cancer (Lynch, 1997; Nakamura et al., 1988; Prows & Brockmeier, 1994).

The FAP gene location has been mapped to chromosome number 5 (Bodmer et al., 1987) and is called the *adenomatous polyposis coli* (APC) gene. The gene defect underlying FAP is believed to be mutation in a tumor suppressor gene that leads to excessive growth and, subsequently, to malignant traits in cells (Schneider, 1994).

Presymptomatic Genetic Testing for FAP: Methodology

Testing for the presence of mutations in the APC gene can be accomplished by direct genetic analysis. This test can identify an APC gene mutation in a patient with FAP in approximately 80 percent of families with a history of FAP. When the precise gene mutation in a family is known, direct testing can differentiate, with nearly 100 percent accuracy, those family members who have FAP from those who do not (Giardello et al., 1997). In some families, direct genetic testing may not be feasible because the specific FAP gene mutation is not known. In these families, linkage analysis using genetic markers may be helpful. Gene markers can be identified in an affected family member, and other family members at increased risk can then be tested for their presence. This determination may be made prior to the onset of symptoms of cancer and can allow for appropriate prevention and treatment interventions. Determination of the presence of the FAP gene can also be performed in a fetus as part of prenatal diagnosis.

Benefits of testing include increased cancer detection and prevention options and resolution of uncertainty. Removal of portions of the colon prior to the onset of malignancy is one option for individuals identified as having the FAP gene. It is currently recommended that, beginning in their late teens, people who are members of FAP families undergo routine endoscopic surveillance. Genetic testing allows for identification of those who have inherited the FAP gene and those who have not and reduces the need for invasive screening for family members who are found not to carry the gene (Bassford & Hauck, 1993; Lynch, 1997; Schneider, 1994).

Current Considerations in Genetic Testing for FAP in Children and Adolescents

Testing of children for FAP prior to the onset of symptoms is an important issue. Current considerations for genetic testing of children and adolescents include the principle that the primary goal of genetic testing should be to promote the well-being of the child; attention to a child's increasing interest and ability to participate in health decisions; and the concept that a child is part of the family network and that decisions about genetic testing must be considered within this context (ASHG/ACMG, 1995; Wertz et al., 1994).

Position statements and recommendations related to genetic testing for cancer susceptibility have been developed by a number of groups (American Society of Clinical Oncology [ASCO], 1996). ASHG/ACMG (1995) recommends that practitioners who provide counseling to families with children at risk for an inherited condition such as FAP include the following activities in the counseling process:

- Assess the significance of the potential benefits and harms of the genetic test.
- Determine the capacity of the child for decision making.
- Advocate on behalf of the best interests of the child.

Education and counseling of families with children being considered for genetic testing is based on appropriate stages of cognitive and moral development. Extensive discussion with the family regarding the potential benefits and harmful effects of genetic testing of their children, including disclosure of test results, is necessary before beginning the testing process. Collaboration with genetics professionals may be a helpful resource to nurses involved in the testing process (Wertz et al., 1994). Suggested components of the counseling and testing process for families and children are provided in Table 6.4.

Additional research focusing on the effectiveness of proposed preventive and therapeutic interventions and the psychological impact of genetic testing on children is ongoing (Andrews et al., 1994; ASHG/ACMG, 1995). As advocates for children, nurses can play an important role in participating and supporting such research efforts. (See Case Study 6.3.)

TABLE 6.4

Components of the Counseling Process for Families and Children

Education and counseling based on appropriate stage of cognitive and moral development

Assurance of informed and voluntary assent of the child in addition to parental permission

Adequate discussion about the potential benefits and harms, and interests of the child when reaching a consensus regarding whether to test

Discussion of disclosure of results to children and adolescents prior to testing

Source: Adapted from the American Society of Human Genetics Board of Directors and the American College of Medical Genetics Board of Directors, ASHG/ACMG report (1995) Points to consider: ethical, legal, and psychosocial implications of genetic testing in children and adolescents. Am. J. Hum. Genet. 57:1233–1241.

CONCLUSION

Genetic testing is rapidly becoming available for many health conditions, and may be offered in a variety of health care settings. The largest categories of such tests include those for prenatal diagnosis of genetic conditions, carrier status, and presymptomatic diagnosis of genetic conditions. Nurses, because they care for patients throughout the life span, will be involved in offering and interpreting tests and supporting those who choose to undergo such genetic testing. To ensure informed consent, nurses will have to be able to provide adequate information to their patients regarding genetic conditions and the testing process. Nurses will also collaborate with the health care team and other specialists during the genetic-testing process. Because of their orientation to family and community care, nurses will be able to address concerns about genetic testing, including the testing of children, and to support all patients in incorporating new genetic information into their daily lives.

SUMMARY POINTS

Genetic testing for carrier, prenatal, and presymptomatic diagnosis, as well as confirmation of a genetic diagnosis, is currently available for many genetic disorders.

Genetic testing methods include direct DNA analysis and indirect methods, such as linkage analysis.

Issues that must be considered when patients are undergoing genetic testing include

- informed consent;
- privacy;
- confidentiality;
- coercion;
- testing when there is no treatment or cure;
- testing of children;
- stigmatization;
- discrimination;
- alteration in self-perception.

Nursing practice activities include

- providing accurate and up-to-date information;
- being aware of social, cultural, economic differences;
- referring to and collaborating with specialized genetic testing centers;
- advocating for children;
- identifying patient and family concerns and coping mechanisms;
- supporting patients and families throughout the testing process.

QUESTIONS FOR CRITICAL THINKING

1. As an office nurse, you encounter a 35-year-old woman who reports a family history of Alzheimer's disease: in her mother, who developed the condition at age 50; her maternal aunt, in whom the disease was diagnosed at age 58; and in her maternal grandfather, who is reported to have died in his sixties from

Alzheimer's disease. The woman explains that she wishes to pursue genetic testing for herself and her two sons, ages 10 and 12. She asks you how she can go about this. Discuss your response. How would you involve a genetics service or genetic-testing center? How would you approach her request to have her sons tested?

2. You are a nurse in a pediatric practice and provide care for a 7-year-old girl who has cystic fibrosis. The parents of your patient are eager to have their two other children, ages 2 and 4, tested to determine whether they are carriers. Discuss the potential issues involved in genetic carrier testing for cystic fibrosis in two children who are siblings of an affected child.

REFERENCES

American Society of Clinical Oncology (1996). Statement of the American Society of Clinical Oncology: genetic testing for cancer susceptibility. *J. Clin. Oncol.* 14:1730–1736.

American Society of Human Genetics Board of Directors and the American College of Medical Genetics Board of Directors (1995). ASHG/ACMG report: points to consider—ethical, legal and psychosocial implications of genetic testing in children and adolescents. *Am. J. Hum. Genet.* 57:1233–1241.

Andrews, L. B., et al. (1994). *Assessing genetic risks: implications for health and social policy.* Committee on Assessing Genetic Risks, Division of Health Sciences Policy, Institute of Medicine. Washington, DC: National Academy Press.

Bassford, T. L., and Hauck, L. (1993). Human Genome Project and cancer: the ethical implications for clinical practice. *Semin. Oncol. Nurs.* 9:134–138.

Billings, P. R., Kohn, M. A., de Cuevas, M., et al. (1992). Discrimination as a consequence of genetic testing. *Am. J. Hum. Genet.* 50:476–482.

Bishop, T. D., and Thomas, H. J. W. (1990). The genetics of colorectal cancer. *Cancer Surg.* 9:585–604.

Bodmer, W. F., Bailey, C., Bodmer, J., et al. (1987). Localizations of the gene for familial adenomatous polyposis on chromosome 5. *Nature* 328: 614–616.

Bulow, S., Svendsen, L. B., and Mellemgaard, A. (1990). Metachronous colorectal carcinoma. *Br. J. Surg.* 77:502–505.

Fitzsimmons, M. (1992). Hereditary colorectal cancers. *Semin. Oncol. Nurs.* 8:252–257.

Giardello, F. M., et al. (1997). The use and interpretation of commercial APC gene testing for familial adenomatous polyposis. *N. Engl. J. Med.* 336: 823–827.

Guidelines for genetic testing for Huntington's disease (1989). *J. Neurol. Sci.* 94:328–332.

Guidelines for genetic testing for Huntington's disease (1990). *J. Med. Genet.* 27:34–38.

Guidelines for the molecular genetics prediction test in Huntington's disease (1994). *J. Med. Genet.* 32:555–559.

Gusella, J. F., and MacDonald, M. E. (1994). Huntington's disease and repeating trinucleotides. *N. Engl. J. Med.* 330(20):1450–1451.

Hagerman, R. J., and Silverman, A. C. (eds.) (1991). *Fragile X syndrome: diagnosis, treatment and research.* Baltimore: Johns Hopkins University Press.

Helix, *National Directory of Diagnostic Laboratories.* Children's Hospital and Medical Center, P. O. Box 5371, CH-94, Sand Point Way NE, Seattle, WA 98105-0371.

Hunt, V., and Walter, F. O. (1991). Learning to live at risk for Huntington's disease. *J. Neurosci. Nurs.* 23(3):179–182.

Huntington's Disease Society of America, Inc. (1994). *Guidelines for genetic testing for Huntington's disease* (revised), pp. 3–13.

Jackson, L. (1987). A predictive test for Huntington's disease. *J. Neurosci. Nurs.* 19(5): 244–250.

Kovach, C. R., and Stearns, S. A. (1993). Understanding Huntington's disease: an overview of symptomatology and nursing care. *Geriatr. Nurs.* 14(5):268–271.

Kremer, B., Goldberg, M. B., Andrew, S. E., et al. (1994). A worldwide study of the Huntington's disease mutation: the sensitivity and specificity of measuring CAG repeats. *N. Engl. J. Med.* 330(20): 1401–1406.

Lessick, M. L. (1993). Pediatric management problems. Fragile X syndrome. *Pediatr. Nurs.* 19(6):622–624.

Lynch, J. (1997). The genetics and natural history of hereditary colon cancer. *Semin. Oncol. Nurs.* 13(2):91–98.

Nakamura, Y., White, R., Smitts, A. M., et al. (1988). Genetic alterations during colorectal-tumor development. *N. Engl. J. Med.* 319:525–532.

Nelson, D. L. (1993). Fragile X syndrome: review and current status. *Growth Genet. Hormones* 9(2):1–15.

Noremolle, A., Sorensen, S. A., Fenger, K., and Hasholt, L. (1995). Correlation between magnitude of CAG repeat length alterations and length of the paternal repeat in paternally inherited Huntington's disease. *Clin. Genet.* 47:113–117.

Oostra, B. A., Jacky, P. B., Brown, W. T., and Rousseau, F. (1993). Guidelines for diagnosis of fragile X syndrome. *J. Med. Genet.* 30(5): 410–413.

Prows, C. A., and Brockmeier, M. J. (1994). Genetic implications of familial adenomatous polyposis: awareness can save lives. *J.W.O.C.N.* 21(5):183–189.

Rimoin, D. L., Connor, J. M., and Pyeritz, R. E. L. (eds.) (1996). *Emery and Rimoin's principles and practice of medical genetics*, 3rd ed. New York: Churchill Livingstone.

Rousseau, F. (1994). The fragile X syndrome: implications of molecular genetics for the clinical syndrome. *Eur. J. Clin. Invest.* 24:1–10.

Scanlon, C., and Fibison, W. (1995). *Managing genetic information: implications for nursing practice.* Washington, DC: American Nurses Association.

Schneider, K. A. (1994). *Counseling about cancer: strategies for genetic counselors.* Dennisport, MA: Graphic Illusions.

Schulman, J. D., Black, S. H., Handyside, A., and Nance, W. E. (1996). Preimplantation genetic testing for Huntington's disease and certain other dominantly inherited disorders. *Clin. Genet.* 49:57–58.

Tarleton, J. C., and Saul, R. A. (1993). Molecular genetic advances in fragile X syndrome. *J. Pediatr.* 122(2):169–185.

Warren, S. T., and Nelson, D. L. (1994). Advances in molecular analysis of fragile X syndrome. *J.A.M.A.* 271(7):536–542.

Warriner, S. (1990). Predictive testing. *Nurs. Times* 86(4):49–50.

Wertz, D. C., Fanos, J. H., and Reilly, P. R. (1994). Genetic testing for children and adolescents: who decides? *J.A.M.A.* 272(11):875–881.

Wexler, N. I. (1990). Presymptomatic testing for Huntington's disease: harbinger of new genetics. In *Genetics, Ethics, and Human Values,* Bankowski and Caprum, eds.

Wilke, J. (1995). From a survivor: the emotional experience of genetic testing. *J. Psychosoc. Nurs. Ment. Health Serv.* 33:28–37.

Williams, J. K., and Lea, D. H. (1995). Applying new technologies: assessment and ethical considerations. *Nurse. Pract.* 20(7):16–26.

FRAGILE X SYNDROME

A 15-year-old boy with a history of cognitive disability and hyperactivity is brought by his parents to their family health clinic for evaluation. At the time of the visit, the mother reports that she is 10 weeks pregnant. In reviewing the family history, the nurse notes that the boy's maternal aunt has two sons who have mental retardation and learning disabilities and who are said to have fragile X syndrome. The parents wonder whether their son has fragile X syndrome. They have questions about whether prenatal diagnosis is available for their current pregnancy. The parents also wonder whether their 7-year-old daughter, who has learning disabilities, may be affected. The mother informs the nurse that she recalls a maternal uncle who was mentally retarded but who is no longer living. She further explains that she has a younger brother who is currently in college and who does not yet have children.

At the clinic, the nurse assists with the physical examination of the boy. The nurse notes that the boy has enlarged testes, larger-than-average ears, delayed speech, and mental retardation, features that are frequently observed in males who have fragile X syndrome. The nurse participates in the discussion with the family about fragile X syndrome. It is explained that medical documentation of fragile X syndrome in the family is helpful in confirming or ruling out the diagnosis. The nurse explains to the family that they will need to ask the mother's sister for permission to obtain the genetic test results that confirm the diagnosis of fragile X syndrome in the mother's two nephews.

The nurse then participates in a discussion with the family about the option of genetic testing of their son for fragile X syndrome and for prenatal diagnosis. The discussion includes a review of the family's questions:

- Does our son have fragile X syndrome?
- Is prenatal diagnosis available for fragile X syndrome?
- Does our daughter also have fragile X syndrome?

The family is referred to a genetics clinic for help in answering their questions and for further evaluation and testing for fragile X syndrome.

At the genetics clinic, a genetic counselor explains that direct genetic testing is available for diagnosis, carrier testing, and prenatal diagnosis of this condition. The counselor reviews with the parents the benefits of testing, specifically that the test could identify the presence of the gene in their son and, if present, in their developing baby and other family members. The counselor also notes that this information would be useful for diagnosing the cause of the son's mental retardation, and could lead to more specific treatments for his condition. The benefits of carrier testing for the daughter are also discussed; among these is the ability to plan for appropriate placement and other special arrangements for their daughter and the ability of the daughter to use this information in future reproductive planning. The counselor informs the family of the limitations of the genetic testing, including the small possibility of noninformative results and the potential for insurance discrimination.

On the basis of their discussion with the genetic counselor and the nurse, the parents decide to pursue genetic testing for fragile X syndrome in their son. A blood sample is obtained from the boy and is sent to a genetics laboratory currently providing testing. Results of the fragile X testing are available in 2 weeks and reveal that the boy has the full mutation known to be present in males who have fragile X syndrome. This information is relayed to the family by the physician in a follow-up visit. The nurse provides the family with written information about fragile X syndrome, including ways to contact a parent support group and the names of a local couple who also have a son with fragile X syndrome and whom they could contact for further supportive information.

At the time of the follow-up visit, the nurse refers the family back to the genetics clinic to discuss prenatal diagnosis for fragile X syndrome. The genetic counselor reviews with the family the risks, benefits,

and limitations of prenatal diagnosis for fragile X syndrome. The counselor explains that direct testing of fetal cells obtained from such prenatal procedures as amniocentesis is available for prenatal diagnosis of fragile X syndrome. Because the diagnosis of fragile X syndrome has already been documented as being present in the mother's son and in her sister's sons, the counselor informs the mother that she is considered to be an obligate carrier of the gene for fragile X syndrome and that she has a 50 percent chance of passing on the fragile X gene mutation to a son or daughter. The counselor further explains that though it is possible, with prenatal diagnosis, to identify males and females who have the full gene mutation for fragile X syndrome, prenatal genetic testing does not always provide information about the degree of severity of fragile X syndrome, especially in females.

The counselor provides the opportunity for further exploration of the mother's feelings about being a carrier and prenatal diagnosis. Reproductive options are also reviewed, including whether to undergo prenatal diagnostic testing, the choices available when there is a prenatal diagnosis of fragile X syndrome, and testing for fragile X syndrome in the newborn period.

The couple decides not to pursue prenatal diagnosis for this pregnancy. The parents explain that their choice is based on the fact that they already have a son with the condition and that it is still not possible to predict the severity of fragile X syndrome in all individuals who inherit the gene. They inform the counselor that, after the birth, they would like their baby to be tested for fragile X syndrome. The counselor notifies the nurse of the parents' decision, and the nurse records this information in the mother's medical record.

In a follow-up visit at the family health clinic, the nurse reviews with the parents the benefits and limitations of testing their 7-year-old daughter and the issue of notifying other family members. Discussion of the potential benefits for the daughter should she be diagnosed with fragile X syndrome include the possibility of more tailored educational support and information that she can use for future reproductive planning. The nurse also raises possible limitations of testing, such as stigmatization and insurance discrimination. The parents decide to wait to pursue testing of their daughter until after their baby is born

The mother informs the nurse that she wants her brother to be tested. She wonders whether the nurse could call her brother's doctor and order fragile X testing for him. The nurse explains to the mother that although it is important to inform other family members of an inherited condition, family members cannot be forced or coerced to accept testing; all individuals have the right to accept or refuse genetic testing. The nurse suggests that the mother contact her brother and offer to send

him information about fragile X testing so that he can make an informed decision that is best for him.

The nurse plans with the family for follow-up throughout the pregnancy. Specifically, the nurse assesses the family's coping mechanisms to support integration of the new genetic information into the family members' daily lives. The nurse makes a plan with the family to continue assessing the impact of the diagnosis of fragile X syndrome on the family.

The nurse also obtains additional information from the genetic counselor about a specialty developmental clinic for individuals who have been diagnosed with fragile X syndrome and about a national support group for families affected by fragile X syndrome. The nurse assures the family that the primary health care provider will arrange for testing of their baby after delivery.

Nursing Issues

Fragile X syndrome is common, so nurses providing primary health care may participate in offering and discussing genetic testing. Knowledge of the clinical aspects of fragile X syndrome will help nurses to identify patients and families who may benefit from genetic testing. Nurses might also collaborate with other health professionals in providing accurate information about the inheritance of fragile X syndrome and the implications of this diagnosis for other family members.

Participating as educators and advocates for informed reproductive choices in discussions with women who are carriers of the fragile X gene are other nursing roles. Nurses in such roles must understand that some women who carry the fragile X gene may also have learning disabilities and varying degrees of mental retardation. A nurse's assessment of a woman's ability to understand complex genetic information is therefore essential to assuring informed consent during the genetic testing process.

A nurse's understanding of the nature of the genetic testing information discussed is equally important. He or she must be aware of how one's personal values might affect the way in which information is presented to a patient. Knowledge of the accuracy of DNA testing for fragile X syndrome, especially with regard to prenatal diagnosis, is also important. Although the options for prenatal diagno-

sis of fragile X syndrome have expanded, some uncertainties still remain, especially with regard to prenatal diagnosis of fragile X syndrome in females. As advocates, nurses can ensure that sufficient information about the prenatal genetic testing process is provided, so that women will feel comfortable about making the best choice for them among the available options.

As educators, nurses can provide support and information to families when a diagnosis of fragile X syndrome is made. Information about genetic services for families are available through local, regional, and national fragile X support organizations. Nurses, in collaboration with genetics and other health professionals, can ensure continuity of care and ascertain that families' needs for information and support are being met.

CASE STUDY 6.2

HUNTINGTON'S DISEASE

Ms. D is a 32-year-old woman admitted to the psychiatric division of a large general hospital because of a history of increasing anxiety and depression. While taking the family and medical history, the nurse learns that Ms. D lives with a 2-year-old daughter and is 16 weeks pregnant. Ms. D

states that the father of her baby is not involved with her at this time. She further explains that her mother, age 55, has HD and is currently living in a nursing home, because she can no longer care for herself. Ms. D also informs the nurse that her maternal grandfather died from HD when he was in his fifties. Ms. D says that she has always worried about the possibility that she might have HD herself but is most concerned about her 2-year-old daughter and whether her developing baby might also have the condition. She asks to speak with someone about testing her daughter for HD and about further information regarding prenatal testing to determine whether her baby has the disease.

The nurse reviews the family history with the medical team and discusses the patient's concerns about HD. The team contacts genetics services to arrange for a genetic-counseling consultation while Ms. D is in the hospital. The nurse coordinates the appointment and reviews with the genetic counselor the family history and the patient's concerns.

In anticipation of the consultation, the nurse describes to the patient the genetic-counseling process. Specifically, the nurse informs Ms. D that genetic counseling is a communication process between the counselor and the patient, the goal being to answer the patient's questions and to offer information about available testing options. The nurse points out that as a routine component of the genetic-counseling process, Ms. D will be asked to provide documentation of the diagnosis of HD in her family.

The nurse accompanies the patient to the genetic-counseling appointment, in the course of which the patient's questions are addressed first. During the counseling session, the nurse learns that the patient has a 50 percent (one in two) chance for inheriting the HD gene from her mother and that prior to testing each of her children has a 25 percent (one in four) chance for inheriting the HD gene. The counselor reviews the testing process with Ms. D and explains the option of having prenatal diagnosis without disclosure of the patient's gene status (Schulman, 1996). Ms. D is also informed of the current guidelines for presymptomatic testing of children and is given written information about HD and available testing for consideration. The counselor explains that she and her colleagues are available to Ms. D and the nursing staff for further discussion, if needed, and will help to arrange for an inquiry call to an HD center if desired.

After the session, the nurse reviews with Ms. D the information about presymptomatic testing and her options. Ms. D informs the nurse that on the basis of what she has learned during the counseling session, she does not feel that she is ready to decide about presymptomatic testing until after she is discharged from the hospital and has had her baby. She recalls that one of her uncles had thought about DNA testing but was concerned about the possibility of losing his job if he tested positive. Ms. D says that she does not want to jeopardize her job opportunities. She states, however, that she is still interested in having her daughter tested, because she is worried about her. The nurse discusses with Ms. D that job discrimination may be an issue. Furthermore, the nurse reviews with Ms.

D the fact that DNA testing of children younger than 18 is not currently recommended because of the issue of informed consent. The nurse offers to help Ms. D arrange a follow-up visit at the HD testing center in a nearby city, so that she can explore these concerns in greater detail with a group of professionals familiar with DNA testing for the disease. The nurse notes this decision in Ms. D's chart and explains to Ms. D that she will receive a letter from the genetic counselor that provides details of their discussion. A copy of this report will also be sent to Ms. D's primary care physician. The nurse provides support for Ms. D's decision and information about local and national HD support groups. The nurse reassures Ms. D by telling her that the nurse and other health care providers are always available for further discussion of any of her health concerns.

The nurse is concerned that Ms. D's depression and anxiety may represent early stages of HD. The health care team is also aware of this possibility and so, in collaboration with the primary care provider, a plan is made for continuous follow-up of Ms. D's condition.

Nursing Issues

Nurses providing health care services to adults in a variety of settings may encounter patients who have HD or a family history of it. Familiarity with the clinical aspects of HD will enhance nurses' understanding of the associated physical and mental disabilities and the implications for an affected individual and his or her family. Nurses can anticipate, for example, that a patient affected with HD will experience alterations in self-concept and in performing their daily living activities. In families living with a member who has HD, coping mechanisms might be altered, as might family members' perceptions of the family unit (Wexler, 1990; Wilke, 1995).

Knowledge of the autosomal-dominant inheritance pattern of HD is important for nursing assessment when eliciting and interpreting family history. Such knowledge will enhance the nurse's ability to identify family members who are at increased risk for inheriting the HD gene. As educators and advocates, nurses can offer information about the inheritance of HD and opportunities for testing, including presymptomatic and prenatal diagnosis (Schulman et al., 1996; Warriner, 1990).

Nurses may be involved in helping patients contact the nearest HD testing center. Knowledge of current recommendations for presymptomatic testing will help nurses to provide appropriate information and support to patients and their families. Nurses may also collaborate with genetics and other specialists in HD testing centers to ensure that a patient has someone to provide emotional support should the patient choose to continue with the testing process and has located a support counselor. Nurses will also need to collaborate with the specialists in HD testing centers to discuss the confidentiality of test results (Hunt & Walter, 1995; Wilke, 1995). Communication and collaboration with local and national HD support groups and additional support through local genetics centers allows nurses to advocate for patients at risk for discrimination (Billings, 1992; Hunt & Walter, 1995; Jackson, 1987; Warriner, 1990).

FAMILIAL ADENOMATOUS POLYPOSIS

At an initial office visit for a company physical examination, John, age 27, informs the nurse of his family history of colon cancer. He explains that his 31-year-old sister has been diagnosed with familial adenomatous polyposis (FAP) and that she recently had surgery. His 50-year-old mother also has a history of FAP, diagnosed when she was in her twenties, and has had surgery. John's mother also had a brother who died from complications of colon cancer when he was in his fifties, and her maternal grandfather is known to have died in his thirties from complications of colon cancer.

John states that he has a daughter, age 10, and two sons, ages 8 and 6. He says that prior to his employment with this company he rarely went to the doctor and therefore does not know whether he has FAP. He says that he has heard that there may be a genetic test to determine whether he has inherited the condition from his mother, and he expresses interest in having the test and in having his children tested.

During the course of the physical examination, the nurse and the physician discuss with John the inheritance of FAP: It is inherited in an autosomal-dominant manner in families; an individual who has FAP has one altered gene for FAP and one normal gene. A parent who has FAP has a one in two (50 percent) chance of passing on the FAP gene to each of his or her children. Because John's mother is known to have FAP, the chance that John inherited the FAP gene is 50 percent.

The nurse explains to John that genetic testing is available for FAP and other types of colon cancer. The nurse also informs John that to consider and pursue genetic testing, it will first be necessary to confirm the diagnosis of FAP in his family, which will require obtaining and reviewing medical records for his mother and sister, with their permission. After the diagnosis of FAP is confirmed in the family, further testing can be offered.

The nurse explains that genetic testing for FAP may involve direct genetic testing of both John and his mother to identify the gene mutation; however, indirect testing for FAP may be used, which requires obtaining blood samples from other family members, such as John's sister and brother. The nurse reviews the benefits of FAP testing, among them identification of the presence or absence of the gene, relief from uncertainty, and when positive, the opportunity to proceed with preventive measures to decrease the risk for developing colon cancer.

At the clinic, John learns that he does not have any signs of FAP at this time. He states that he nevertheless wishes to proceed with testing.

John's mother and sister give permission for review of their medical records, which document a diagnosis of FAP in the family. In a follow-up visit with John, the nurse and physician review the diagnosis of FAP in the family and the testing process. The nurse explains to John the informed-consent form provided by the laboratory. Specifically, she reviews the accuracy of the current testing methods and the limitations of the testing, including the possibilities of noninformative results, test error, and nonpaternity.

The physician and nurse discuss with John the benefits of testing, including cancer detection and prevention options, resolution of uncertainty, the option to test John's children if John is found to have the gene, and the increased ability to plan for the family's future medical needs. The possible adverse outcomes of learning certain genetic information are also explained; among these possible outcomes are knowledge of an increased cancer risk, anxiety and lowered self-esteem, strained relationships with other family members, feelings of guilt when results are negative, possible loss of insurance and discrimination by employers, and the costs of extra cancer surveillance.

After careful consideration with his family, John decides to pursue testing. The nurse arranges for a consultation with the genetics and cancer specialists, who collaborate with John's physician to plan testing for John and his family. Preliminary genetic testing and analysis reveal that a specific FAP gene mutation has not been identified in John's family and therefore direct testing for FAP will not be possible. Linkage analysis is therefore offered, and John's sister, mother, and brother are informed of the testing process. Because they live in a different state, the nurse identifies for these family members a local genetics and testing center and helps arrange for genetic counseling and coordination of family testing.

Results of the tests are available in 3 weeks. The nurse receives notification from the genetic testing laboratory that John does carry linkage markers for the gene for FAP. The diagnosis of FAP is confirmed in John's mother and sister, though John's brother is found not to have inherited the linkage markers for the FAP gene.

John is informed of his test results. He asks to have his children tested, so that they will receive proper cancer surveillance if needed. The nurse explains that it is important for the children to be included in the informed-consent process, even though they are young; explains the importance of assuring that they understand the meaning of the testing for themselves; and informs John that genetics professionals, including a social worker familiar with the genetic testing process, are available to the family for further counseling.

John and his children seek additional information and counseling from the genetics professionals. After counseling and discussion about the benefits of testing the children to identify those who need close surveillance, each of them agrees to be tested. Test results show that John's daughter and younger son do not have the FAP gene; however, his 8-year-old son has inherited it. Follow-up counseling is arranged with the son's pediatrician and social worker to develop a plan for surveillance for both John and his son. The nurse ensures that counseling is also made available for the daughter and youngest son.

The nurse's interventions include follow-up assessments of the family with the social worker, in anticipation of any adverse consequences of genetic testing. The nurse ensures that the family is offered additional information about local and national FAP support groups.

Nursing Issues

Nurses will become increasingly involved in assessing families for inherited forms of colon cancer such as FAP and in offering, discussing, and interpreting genetic testing for FAP. Presymptomatic or prenatal testing for APC gene mutations requires informed consent. The nurse fills the important role, when recording the family history, of identifying families at increased risk and assuring that those patients and families are provided with full information about the nature of the condition, currently recommended methods of surveillance, and available genetic testing (Prows and Brockmeier, 1994). This role includes informing a client that testing for FAP will usually provide information that will help him or her to make informed health decisions; however, in some families, the specific mutation in the family may not be known, and linked markers may prove to be uninformative, so that the testing may not be helpful.

Once a patient has been diagnosed with FAP, his or her entire family must be offered information regarding both their chances for having inherited the FAP gene and available medical and genetic testing (Fitzsimmons, 1992; Lynch, 1997). The nurse may collaborate with genetics counseling and testing centers, when family members are not from the same geographical area, to arrange for family members to learn about available counseling and genetic-testing opportunities. The nurse can also help patients to consider some of the issues that may arise during the genetic-testing process, such as the impact of the diagnosis of FAP on themselves and other family members. As an educator and advocate for informed health decisions, the nurse can provide information that will assist patients in discussing FAP and available testing with other family members.

Knowledge of the inheritance of FAP, the concept of risk, and the potential outcomes of testing prepares nurses to provide appropriate support for family decision making. Primary-care nurses may collaborate with genetic nurse specialists and counselors to prepare for discussion of these issues. Emotional support is an ongoing nursing intervention that is offered to patients and families as they move through the decision-making and genetic-testing processes. Supportive nursing care involves the ability to articulate to patients and families that each person has the right to choose or to refuse genetic testing and the ability to impress upon people that these are personal choices.

Assurance of the privacy of each family member and of the confidentiality of test results are of central importance in offering genetic testing (Scanlon & Fibison, 1995; Schneider, 1994; Warriner, 1990). Results of genetic testing for FAP are likely to be of interest to other family members, such as aunts, uncles, or cousins, who may wish to pursue testing for themselves. In such situations, the nurse, as an advocate for confidentiality and privacy, ensures that before information is given to others, permission is obtained from each family member in whom the diagnosis is made or who undergoes genetic testing. Protection of the privacy of genetic information is especially important in families in which several members are involved in FAP testing and counseling. The nurse has an important role in maintaining the confidentiality of each person's genetic information (Scanlon & Fibison, 1995).

Throughout the counseling and testing process, the nurse must be aware of the potential for altered self-esteem in family members who learn about the presence or absence of FAP. Results of studies of patients who have had genetic testing for other conditions, such as Huntington's disease, sickle-cell anemia, and Tay-Sachs disease, have shown that depression, anxiety, concern about one's health, shame, and guilt may be associated with learning about one's gene status (Billings et al., 1992; Marteau, 1992; Sharpe, 1994; Wilfond & Fost, 1990). Continuous assessment of the patient's emotional status and his or her understanding and interpretation of genetic test results is an important nursing activity throughout the counseling and testing process. Collaboration with other health professionals, such as genetic nurse specialists, counselors, and social workers, may also support the nurse in providing care for these individuals and families.

The possibility of discrimination based on genetic test results is often of concern to patients. Denial of reimbursement for genetic testing, an increase in insurance rates, and loss of eligibility for insurance have been documented in some cases for patients and families who have chosen to pursue genetic testing for various inherited conditions (Billings et al., 1992; Schneider, 1994; Andrews et al., 1994). This aspect of genetic testing must be discussed during the decision-making process (Williams & Lea, 1995).

The Genetic Basis of Cancer

Most people don't really learn anything, because they're still thinking about today's problems. Today's problems are already solved. What I want to do is open up new areas, and once it opens up and other people get into it, I'm happy.

W. French Anderson

OBJECTIVE

Demonstrate increased knowledge of cancer genetics and current applications to practice.

RATIONALE

Nurses will be involved with patients seeking genetic information, undergoing cancer-risk assessment, and experiencing genetic testing as a part of general cancer health care. Nurses will therefore need to have a basic understanding of cancer genetics in order to identify individuals and families who are in need of additional information, referral, and testing, and to provide support and guidance to patients and families throughout this process.

APPLICATION ACTIVITY

Provide a brief explanation that genetic changes are present in all cancers but that not all cancer is inherited.

GENE CARE LINK

For access to information about cancer genetics resources, go to http://www. jbpub.com/ clinical-genetics.

INTRODUCTION

enetic discoveries are increasing our understanding of the cause of cancer and are contributing to new ways to diagnose, monitor, and treat cancer. This chapter provides an overview of both our current understanding of the genetic contribution to cancer and of approaches to assessing and counseling patients and families who are at risk for cancer. Nurses may adopt many of the components of risk assessment, counseling, and intervention in all patient-care settings in the future, as genetics becomes integrated into nursing practice. Oncology nurses are likely to be among the first to have this opportunity and to model and assess outcomes of nursing care that integrate into patient care the principles of genetic knowledge, risk assessment, and counseling (Olufunmilayo & Cummings I & II, 1996).

THE GENETIC BASIS OF CANCER

Cancer is a group of diseases in which the regulation and the maturity of normal cells is disturbed. Numerous types of cancer exist, and even cancers of the same type can behave very differently from each other. Despite this diversity, certain fundamental defects are common to all types of cancer. As Nobel Laureate Harold Varmus (Varmus & Weinberg, 1993) has noted, "Cancer cells divide without restraint, cross boundaries they were meant to respect, and fail to display the characteristics of the cell lineage from which they were derived." These changes in growth and development are under genetic control. Cancer is, therefore, a genetic disease at the cellular level. This does not mean, however, that all cancer is inherited. It *does* mean that cancer develops when crucial genes are damaged in ways that cause cells to escape their normal controls and begin to grow and divide chaotically.

Uncontrolled cellular growth distinguishes a cancer cell from a normal cell and occurs when a gene or genes that usually control this growth are altered (mutated) in a cell. This alteration results in dys-regulated cell function, such that uncontrolled growth is initiated and malignancy occurs (Figure 7.1). Three properties that distinguish a cancer cell from a normal cell are *immortalization, transformation,* and *metastasis* (Lewin, 1994). Immortalization describes a cell's ability to grow through an indefinite number of divisions, transforming into cells that grow in an unrestricted manner. These changes permit cancer cells to invade normal tissue and to spread, or metastasize, throughout the body. This process, which results in abnormal cell expression characteristic of malignancy, is called *carcinogenesis.*

FIGURE 7.1

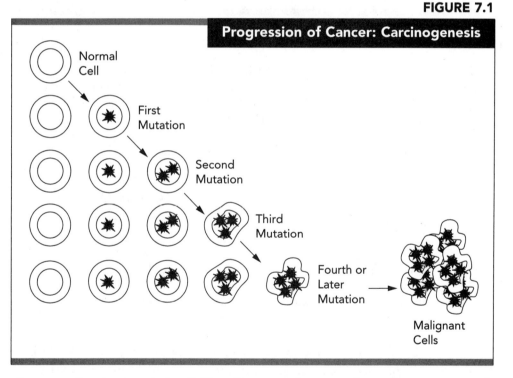

Cancer usually arises in a single cell. The cell's progress from normal to malignant to metastatic appears to follow a series of distinct steps, each controlled by a different gene or set of genes. People with hereditary cancer already have the first mutation. (Adapted from National Institutes of Health and National Cancer Institute, Understanding Gene Testing [NIH pub. no. 96-3905]. Washington, DC: U.S. Department of Health and Human Services, 1995.)

Oncogenesis

Cancer rarely occurs as a result of one event. Rather, oncogenesis is believed to be a multistep process resulting from the accumulation of numerous genetic mutations that cause uncontrolled cellular growth. Tumor-cell DNA undergoes one or more mutations that alter, eliminate, or dysregulate the cell as a result of genetic changes. These mutations in the genetic material may be point mutations, deletions, insertions, amplifications, or translocations. (See Table 7.1 for clarification.) The mutations cause the cell to become progressively dysregulated in its growth, leading to malignant transformation.

TABLE 7.1

Gene Mutations That May Cause Cancer

Gene mutations involve a permanent heritable change in the sequence of DNA. Changes in the normal sequence of bases can alter the characteristics of the proteins for which DNA codes. Following are examples of various types of mutations using a sequence of three words: *CAT ATE RAT.*

Normal base sequence	CAT ATE RAT
Point mutation	CAR ATE RAT
Deletion	CAT AT RAT
Insertion	CAT ATE RATS
Amplification	CATTTTTT ATEEE RATTTTT
Translocation	RAT ATE CAT

Normal DNA replication is not 100 percent precise under the best of conditions. Nonetheless, the body has an intricate system of checks and balances that normally corrects or repairs such genetic mutations. Disruptions in these corrective and reparative mechanisms happen naturally with aging, which explains the increased prevalence of cancers in late adulthood. Genetic information can also be altered through exposure to carcinogens (agents with the potential to initiate or cause cancer), through inheritance, or by chance. Genetic predisposition to the effects of carcinogens and other environmental influences depends not only on the intrinsic properties, timing, and dose of the agent but also on the host's responses.

Many processes that maintain the fidelity of DNA replication and repair become increasingly erratic in cancer cells. Some of these changes may help the premalignant cell to grow more rapidly, to compete more effectively with normal cells for essential nutrients, to relax the control mechanisms of cell differentiation, and to progress toward a more malignant type of cell. The ways that genetic information changes are not mutually exclusive. The multistep process of carcinogenesis, in fact, requires multiple cumulative genetic changes to occur over time. Studies of different types of cancer reveal that there are a variety of genetic changes implicated in the development of cancer. This section discusses the many ways in which these changes can occur.

CAUSES OF ONCOGENESIS

Acquired

Most cancers are caused by multiple factors (multifactorial). Noninherited factors, such as environmental exposures, and inherited factors interact and predispose a person to develop cancer. Noninherited factors, including chemical, physical, or biological exposures, cause *somatic mutations* in a cell. Somatic mutations are gene alterations in individual cells that may not be repaired and are incorporated into all future descendants of the altered cell. Somatic mutations occur and accumulate as a result of exposures during a lifetime, but they are not passed on to offspring because they are not present in reproductive cells (oocytes and sperm).

Inherited

Nearly all types of cancer have been reported to occur in families. Familial cancers account for approximately 5–10 percent of total cancers and are due primarily to inherited gene mutations (Fraser et al., 1997). *Inherited gene mutations* (germline mutations), in contrast to somatic mutations, are those that are present in the DNA of every cell of a person's body, including oocytes and sperm. An inherited, or germline, mutation, therefore, does have the potential to be passed on to offspring. Patients who inherit cancer-susceptibility genes develop cancer at a younger age than does the average person (Weber et al., 1996). Patients who have cancer-susceptibility genes may experience further adverse effects when exposed to environmental factors such as chemical, radiation, or biological agents known or suspected to be carcinogenic.

Hereditary nonpolyposis colon cancer (HNPCC) is an example of an autosomal-dominant inherited cancer condition. The risk of colon cancer in a person who has inherited mutation in one of the HNPCC genes is greatly increased. In some affected families, the risk of developing other types of cancers may also be increased.

Other inherited Mendelian conditions are associated with an increased risk (predisposition) of cancer. Patients who have neurofibromatosis type 1, for example, are at increased risk for developing brain tumors, whereas individuals in families with the Li-Fraumeni syndrome may develop early-onset leukemia or breast or bone cancer. Both neurofibromatosis type 1 and the Li-Fraumeni syndrome are inherited in families in an autosomal-dominant manner.

CLASSES OF CANCER-CAUSING GENES

Several classes of genes have been identified as regulating or inhibiting cell growth and are thus believed to contribute to cellular transformation that results in cancer. These genes include oncogenes, tumor-suppressor genes, mismatch repair genes, apoptosis genes, and telomerase genes.

Oncogenes

An oncogene is a cancer-causing gene carried by an acute transforming retrovirus that has a normal cellular counterpart called a *proto-oncogene*. The proto-oncogene is activated to become an oncogene. Cancer-causing oncogenes were identified and named according to the virus in which they were originally carried.

The generic term *oncogene* is now used to describe any gene linked to carcinogenesis or uncontrolled cellular growth. An oncogene can be thought of as the accelerator in a car. Its role in cancer development is that of promoting uncontrolled cellular growth. Oncogenes behave in a dominant fashion, which means that a single mutation in one of a gene pair is sufficient to promote carcinogenesis. Approximately 100 oncogenes have been identified, including those oncogenes associated with pediatric leukemias, lymphomas, and various solid tumors. Current cancer research is focused on assessing the presence of oncogenes in diagnosing cancer, predicting clinical outcomes, enhancing therapeutic options, and targeting treatments.

Tumor-Suppressor Genes

Tumor-suppressor genes are cellular growth inhibitors and have been compared to the brakes of a car. Tumor-suppressor genes make proteins that restrain cell growth and prevent cells from becoming malignant. The presence of even one functioning allele of a gene pair can often accomplish this task and prevent the development of cancer. People born with one absent or nonfunctional allele of a gene pair are at much higher risk for developing cancer because they have only one functioning allele in reserve. If the second allele is inactivated, then cellular growth is no longer well controlled. Tumor-suppressor genes function in a recessive manner, meaning that both alleles of the gene must be altered in order for the cellular function to be lost. Inactivation of tumor suppressor genes is now believed to contribute to many cancers. Mutations in the p53 tumor-suppressor gene, for example, have been implicated in 50 percent of all cancers (Karp, 1996).

Mismatch Repair Genes, Replication Errors, and Cancer

Mutations in genes that control the accuracy of DNA replication during cell division have been implicated in cancer development. These genes, called *mismatch repair genes,* are responsible for overseeing DNA replication and for repairing any mistakes that occur during the replication process (Holzman, 1996). An alteration in the mismatch repair genes, similar to a defective spell-checking mechanism on a computer, prevents both the detection and correction of errors as DNA is copied. Gene mutations in mismatch repair genes may be inherited and have been identified in Hereditary Non-Polyposis Colon Cancer (HNPCC), a condition characterized by colorectal, endometrial, and other cancers.

Apoptosis Genes

Cells grow old and die by activating a program that leads to their death and are then deleted for the overall good of the organism. This death pathway is called *apoptosis* (Sluyser, 1996). Disruption of this pathway can lead to cancer when cells that would normally die continue to live and thus continue to collect mutations that make them more susceptible to malignant transformation.

Telomerase (Aging) Genes

Multiple genes are involved in the control of cell aging and death. One product, the enzyme telomerase, is being more closely evaluated for its role in the cellular aging and immortalization mechanism. Telomeres are the ends of chromosomes. Telomerase, which helps to maintain telomeres, is present during fetal development and then becomes repressed in somatic cells after birth. As somatic cells age, small pieces at the ends of chromosomes are lost, a process that leads to a loss of information and loss of the telomeres. A current theory is that telomerase is reactivated in cancer, preventing telomere erosion and allowing cell immortalization; a gene mutation in telomerase allows cells to continue to proliferate and divide and leads to cancer. Antitelomerase therapy is being considered as a future treatment for some cancers (Look & Kirsch, 1997; Shay & Wright, 1996).

CHROMOSOMAL AND GENETIC ABNORMALITIES IN CANCER

Chromosomal Abnormalities

Many types of cancer feature an abnormal number or arrangement of chromosomes. Chromosomal abnormalities resulting from either more or fewer than the normal 46 chromosomes are called *aneuploidy* and are manifestations of genetic instability. This instability can cause an increased risk for cancer. Patients who have Down syndrome, for example, are at increased risk for developing leukemia.

Structural abnormalities of chromosomes that cause an increased risk for cancer occur when a large segment of genetic material located on one chromosome is inverted, deleted, duplicated, or transferred to a place on or between chromosomes where it isn't normally found. Chromosomal rearrangements of this type frequently result in the formation or activation of oncogenes in human tumors. Leukemia cells, for example, often show rearrangements, duplications, and deletions of portions of their chromosomes during cell division. The result is a new combination of genetic material. This rearrangement may be used to make the diagnosis of a specific type of leukemia, such as the Philadelphia chromosome (a chromosome rearrangement between chromosomes 9 and 22) seen in chronic myelogenous leukemia.

Viruses

Insertion of a viral oncogene (a cancer-causing gene) in a normal cell is a way in which the genetic information of a normal cell can be disrupted. Viruses often kill the cells that they infect, but sometimes a cell survives infection by certain viruses and viral genetic information is integrated, which can cause the cell to become cancerous. The Epstein-Barr virus, associated with the development of Burkitt's lymphoma, is such a virus. The hepatitis B virus, which causes liver cancer, and certain strains of human papillomavirus, which contribute to the development of cervical cancer, are other examples. Patients infected with the human immunodeficiency virus (HIV) experience a higher incidence of cancer as a result of this mechanism of cell infection. These patients' depleted immune systems may promote the development of conditions such as Kaposi's sarcoma, lymphoma, and anogenital cancers.

Gene Amplification

Gene amplification is increased gene expression and can be thought of as "turning up the volume" of a gene message; it occurs when an error in DNA replication leads to the presence of multiple copies of an oncogene which gives a cell extra messages to grow. Cells containing the amplified oncogene undergo a quantitative or qualitative change in their gene products, which produces an advantage over surrounding cells and contributes to aggressive growth of the malignancy. The amplification of an oncogene in some stages of the cancer (e.g., in neuroblastoma) is an example of this process.

Gene Mutations

The sequence of nucleotide base pairs in DNA makes up the genetic code. Any change in this sequence is called a *mutation*. The bases of the genetic code are transcribed from DNA to RNA in groups of three called *codons*. Each codon in the RNA, after some editing, is translated into an amino acid that produces the final protein product. A gene mutation causes an abnormal RNA which, in turn, may result in an abormal protein (Korf, 1996). This process was described in more detail in Chapter 3.

Different types of gene mutations that contribute to the development of cancer have been identified. They are classified as *nonsense mutations, missense mutations,* and *frameshift mutations* (Table 7.2).

TABLE 7.2

Definitions of Mutations

Nonsense mutation: a point mutation that creates a premature "stop" codon, which shortens the gene product.
    Normal: CAT ATE RAT
    Nonsense: CAT ATE

Frameshift mutation: deletion or addition of one, two, or any number of bases that is not a multiple of three. This disrupts the normal reading frame, resulting in premature termination of translation.
    Normal: CAT ATE RAT
    Frameshift: CA TAT ERA T

Missense mutation: a single base change that alters an amino acid in the gene product, resulting in an abnormal protein.
    Normal: CAT ATE RAT
    Missense: CAT ATE RAG

A missense mutation is caused by an alteration in a single base of a triplet of bases that codes for a specific amino acid. When a missense mutation occurs, a different amino acid is formed and incorporated into the developing protein. If the newly formed amino acid is functional, then the protein may function adequately. This type of mutation is said to be benign and does not contribute to cancer. If, on the other hand, the mutation occurs at a critical site in the gene, then it can lead to an altered protein that is not able to function properly and may cause cancer.

Either a nonsense or a frameshift mutation may occur during translocation of DNA, and this affects the expression or function of the protein product. A nonsense mutation can cause unstable proteins to form as a result of a nucleotide substitution. A frameshift mutation is one that may cause DNA transcription to stop prematurely or that may cause a short or abnormally configured protein to be formed. These types of mutations appear commonly in cancer and have a role in activating cancer-causing genes or inactivating genes involved in cancer regulation.

Gene Deletions

Gene deletions are often seen in cancer tumors. In this situation, a single copy of a gene may be mutated. If the second copy of the same gene is lost due to a gene deletion, the cell has no working copy of that gene. This type of mutation is typical of abnormalities in tumor-suppressor genes. The brain cancer glioblastoma is an example of cancer due to a deleted copy of a tumor-suppressor gene.

Increased Chromosomal Breakage and Defective DNA Repair

A variety of inherited conditions are characterized by abnormalities in chromosomal repair mechanisms and by DNA damage. These abnormalities are associated with an increased risk for congenital defects as well as for certain types of cancer. Defective DNA repair occurs, for example, in ataxia telangiectasia, a condition in which patients cannot repair radiation damage to DNA. Conditions such as Bloom's syndrome involve a mutation in a gene that codes for a protein necessary for maintaining chromosome stability. Both ataxia telangiectasia and Bloom's syndrome are examples of inherited conditions that create a predisposition to developing cancers in childhood or early adulthood as a result of genetic changes.

Imprinting and Cancer

Genetic material can also be altered by a phenomenon called *genomic imprinting*. Imprinting occurs when a specific oocyte or sperm is altered in a way that causes a difference in the expression of the two alleles of a gene in the somatic cells (Feinberg, 1993). Some genes are chemically marked, or imprinted, causing them to be expressed differently, depending on whether they are inherited from the mother or the father. Imprinting has been documented in several types of cancer, including Wilms' tumor, bilateral retinoblastoma, and neurofibromatosis type 1.

THE FUTURE OF CANCER GENETICS

Our ability to learn more about the underlying mechanisms of cancer is increasing as more is understood about genetic influences on growth and development and about the intricacies of the checks-and-balances system in the human body. Technological advances are helping us to discern whether cancer results from a single genetic change, from a cascade of events, or as a result of years of environmental exposures. Future discoveries regarding the genetic contribution to cancer development, progress, and responsiveness to interventions are expected to continue at a rapid pace (Gottesman, 1994). The challenges ahead for nurses and other health care providers include preparation for integrating this new knowledge into clinical practice.

The National Cancer Institute (NCI) is laying the foundation for a new era of cancer diagnosis using molecular techniques (Kuska, 1996). Through an organization called the Genome Anatomy Project, the NCI is seeking and sharing information about genes expressed in cancer. The goal of this project is to apply knowledge of new technology and biomarkers to clinical diagnosis and treatment of cancer. A cancer-genetics network has been funded to build an infrastructure of research and testing sites by which to link genetic testing to research and intervention trials (Nelson, 1997). It is expected that the NCI will provide funding for cancer centers around the United States to meet the needs of patients and families for cancer-genetics services. Most cancer centers are currently in the developmental phases of planning, educating staff, and preparing to provide specialized clinical services for familial cancer (Thompson et al., 1995).

CONCLUSION

Knowledge of the genetic aspects of cancer is opening up new avenues for clinical risk assessment, diagnosis, and treatment of patients and families. The ability to identify patients and families who have a genetic predisposition to develop cancer offers the potential for earlier discovery and treatment of cancer. New molecular technologies are allowing measurement of factors that influence cancer development and will help us to monitor patients' response to exposure to such factors.

The finding that many types of cancer share the same genetic defects raises the hope that many cancers may be treated using the same approach. Further clinical trials will help to evaluate the outcomes of cancer therapies directed at specific gene mutations. Clinical trials to assess the efficacy and safety of gene therapy are ongoing (Roth & Cristiano, 1997). Gene-therapy methods require locating the mutated gene, determining how it functions, and identifying reliable ways to administer the corrected gene to the patient safely and effectively. Other trials that are under way involve using prognostic markers to assess drug sensitivity or resistance in cancer and to track responsiveness of cancer to treatment. One goal of these trials is to determine individually designed cancer-treatment regimens. The success of these new diagnostic and treatment options depends on solving many conceptual and technical problems. The challenge is finding ways to apply molecular and biochemical research in clinical practice to enhance the outcome for patients with cancer.

Summary Points

Cancer

- a genetic disease at the cellular level
- not always inherited
- a multistep process
- characterized by immortalization, transformation of cells, and metastasis

Causes of cancer

- acquired
- inherited

Cancer-causing genes

- oncogenes
- tumor-suppressor genes
- mismatch-repair genes
- apoptosis genes
- telomerase (aging) genes

Chromosomal and genetic abnormalities

- change in the normal sequence of DNA
- point-base mutation
- deletion
- insertion
- amplification
- translocation

Future directions

- enhanced understanding of the genetic contribution to cancer development, progress, and responsiveness to intervention
- Cancer Genome Anatomy Project
- cancer-center involvement
- clinical trials to address gaps in knowledge

QUESTIONS FOR CRITICAL THINKING

1. If you were asked to describe to a patient with an eighth-grade educational background the genetic basis of cancer, how would you do so? What educational material might you use?

2. Describe the difference between somatic and germline gene mutations.

3. Describe components of nursing practice in which the integration of cancer genetics into the nursing role might occur.

REFERENCES

Feinberg, A. (1993). Genomic imprinting and gene activation in cancer. *Nature Genet.* 6:273–281.

Fraser, M., Calzone, K., and Goldstein, A. (1997). Familial Cancers: new and evolving clinical issues. In Hubbard, S. M., Goodman, M., and Knobf, M. T., *Oncol. Nurs. Updates* 14(3). Cedar Knolls, NJ: Lippincott-Raven.

Gottesman, M. (1994). Report of a meeting: molecular basis of cancer therapy. *J. Natl. Cancer Inst.* 86:1277–1285.

Holzman, D. (1996). Mismatch repair genes matched to several new roles in cancer. *J. Natl. Cancer Inst.* 88:950–951.

Karp, G. (1996). *Cell and molecular biology.* New York: Wiley.

Korf, B. (1996). *Human genetics.* Boston: Blackwell Science.

Kuska, B. (1996). Cancer genome anatomy project set for take-off. *J. Natl. Cancer Inst.* 88:1801–1803.

Lewin, B. (1994). *Genes V.* New York: Oxford University Press.

Look, A., and Kirsch, I. (1997). Molecular basis of childhood cancer. In P. Pizzo and D. Poplack (eds.). *Principles and practice of pediatric oncology* (3rd ed.), pp 37–74. Philadelphia: Lippincott-Raven.

National Institutes of Health, National Cancer Institute (1995). *Understanding gene testing* (NIH pub. no. 96-3905). Washington, DC: U.S. Department of Health and Human Services.

Nelson, N. (1997). Cancer genetics network gets underway with 5 years of funding. *J. Natl. Cancer Inst.* 89:10–12.

Olufunmilayo, O., and Cummings, S. (1996). Genetic counseling for cancer: part I. *Princ. Pract. Oncol.* 10:1–12.

Olufunmilayo, O., and Cummings, S. (1996). Genetic counseling for cancer: part II. *Princ. Pract. Oncol.* 10:1–11.

Roth, J., and Cristiano, R. (1997). Gene therapy for cancer: what have we done and where are we going? *J. Natl. Cancer Inst.* 89:21–39.

Shay, J., and Wright, W. (1996). Telomerase activity in human cancer. *Curr. Opin. Oncol.* 8:66–71.

Sluyser, M. (ed.) (1996). Apoptosis in normal development and cancer. Philadelphia: Taylor and Francis.

Thompson, J., Wiesner, G., Sellers, T., et al. (1995). Genetic services for familial cancer patients: a survey of National Cancer Institute Cancer Centers. *J. Natl. Cancer Inst.* 87:1446–1455.

Varmus, H., and Weinberg, R. (1993). *Genes and the biology of cancer.* New York: Scientific American Library.

Weber, W., Mulvihill, J., and Narod, S. (1996). *Familial cancer management.* New York: CRC Press.

Cancer-Risk Assessment and Counseling

What do we live for, if it is not to make life less difficult for each other?
George Eliot

OBJECTIVES
- Describe the essential elements of cancer-risk assessment.
- Describe the essential elements of genetic-susceptibility testing.
- Describe the essential elements of emerging nursing roles.

RATIONALE
Nurses' familiarity with the elements of cancer-risk assessment and genetic-susceptibility testing will ensure that nurses carry out these activities knowledgeably and sensitively. The opportunity to integrate concepts of cancer genetics into practice will present first to oncology nurses, who can then provide a role model for others.

APPLICATION ACTIVITIES
- Identify individuals and families in need of further cancer-risk assessment and testing.
- Refer patients and families to specialized testing centers and support groups.
- Provide anticipatory guidance to families seeking cancer-risk assessment and testing.
- Participate in screening of patients, using family history and physical assessment.
- Collaborate with cancer-risk assessment and intervention teams.
- Provide support and follow-up throughout the counseling and testing process.
- Monitor patients' responses to cancer diagnostic, preventive, and intervention methods.

GENE CARE LINK

For educational resources for teaching about cancer genetics, go to http://www. jbpub.com/ clinical-genetics.

193

INTRODUCTION

Each encounter between a nurse and a patient provides an opportunity for evaluation of genetic information, such as the patient's health history. Information about potential risk factors—environmental exposures or a family history of cancer—is easily obtained during a patient's visit. This information can then be used to identify which patients and families may benefit from further risk assessment and counseling for cancer. Nurses can also identify patients and families at increased risk for cancer through medical-record review, physical assessment, and laboratory studies, and through listening to patients' perceptions of risk. A nursing assessment for cancer risk includes questions regarding the family history: type of cancer, age at diagnosis, outcome of illness, and age at death of affected family members. Gathering this information also assists in identifying patients and families who are at increased risk for cancer susceptibility. Knowledge of the genetic contribution to cancer and the components of risk assessment and counseling for cancer will help nurses to participate more fully in educating, counseling, and interpreting for patients information regarding the causes of cancer and the testing and treatment options.

COMPONENTS OF CANCER-RISK ASSESSMENT: A CASE EXAMPLE

Throughout this chapter, the case of WR will be used to illustrate the concepts presented herein.

WR, a 43-year-old man who is experiencing bleeding and pain with bowel movements, is accompanied by his wife for a visit to a primary care physician. WR has suffered some abdominal discomfort, distention, and nausea over the last few weeks but attributed these symptoms to the flu. The nurse takes the opportunity to perform an admission assessment, which includes questions about WR's work history, lifestyle, and dietary habits and family experiences with illness.

Identification and Screening

The purpose of risk assessment for cancer is to obtain individual patient information about the probability of that patient's risk for developing cancer. Identification or case finding takes into account a person's lifetime experience with cancer, genetic predisposition (as indicated by family history), environmental exposures, lifestyle influences, socioeconomic and ethnic backgrounds, and personal risk perception.

The nurse notes that WR has a history of colon cancer on his mother's side of the family, which raises the nurse's level of concern about WR's symptoms. WR also reports that he had a benign polyp removed from his colon 2 years ago but has not had any other problems since that procedure.

Analysis of Risk Factors

Although the variables suggestive of a hereditary cancer predisposition are continuously being determined, some features are currently considered risk factors, or "red flags." These features include increased frequency of cancer in the family, early age at diagnosis (younger than 50), cancer in paired organs, or unique tumor combinations. See Table 8.1 for more details. Using a pedigree to diagram the family relationships and incidence of disease can help to clarify the potential for increased risk of cancer development (see Chapter 3 for more details on pedigree construction).

WR briefly tells the nurse his family medical history: His mother was diagnosed with colon cancer at age 49 and is still living. Her brother and father have died from colon cancer. WR has a brother, age 35, who is healthy, and a sister who had ovarian cancer, diagnosed at age 35. Having formed the opinion that this family may be at increased risk for an inherited predisposition to cancer, the nurse notes the assessment in the chart and shares this information with the family physician.

TABLE 8.1

Assessment for Cancer-Risk Factors

Frequency	Two or more relatives of one generation, or three or more relatives in two generations, having the same cancer
Age at Diagnosis	Cancer development usually 10 to 15 years earlier than is commonly found for a given cancer
Cancer presentation	Often bilateral; possibly multiple primary tumors; unique histology or site occurrence; may be associated with precursor lesions or found within well-defined hereditary syndromes such as familial adenomatous polyposis, retinoblastoma, or von Hippel-Lindau syndrome

Source: Modified from Fraser, Calzone, and Goldstein, 1997.

Risk Analysis

Once identified, patients at increased risk are usually referred to specialized clinics for further risk analysis. In such practices, models that assist clinicians in predicting statistical risk are often used. Examples of statistical models include the Gail and Claus models for breast-cancer risk. The Gail model (Gail et al., 1989) is multifactorial in approach (age at menarche, number of pregnancies, etc.), whereas the Claus model (Claus et al., 1994) uses number, degree of relationship, and age at cancer diagnosis of relatives to predict risk. The advantage of using a risk-modeling system is that it provides the patient with some estimate of his or her risk. However, the disadvantage is that none of the current models are considered fully reliable in high-risk patient populations. Work on such models continues for further validation and use in patient care.

The family physician is immediately alarmed by WR's family history and by the report of the patient's previous polyp removal. The physician concentrated on the immediate concern of diagnosing the source of WR's rectal bleeding and other symptoms. WR underwent a colonoscopy, which revealed another polyp. A biopsy of the polyp showed it to be cancerous.

DIAGNOSING HEREDITARY CANCER SYNDROMES

A diagnosis of a genetic syndrome is based on the number of affected and unaffected relatives, the relationships of these relatives to one another, recognition of a constellation of characteristic problems for that specific cancer, and age at onset of symptoms. As with any genetic condition, an accurate diagnosis allows for more complete genetic counseling of the family. Every effort is therefore made to establish a diagnosis.

Patients who have a family history of hereditary cancer generally have two medical concerns: the chance that they will inherit a cancer susceptibility gene mutation, and the chance, if they have inherited a cancer-susceptibility gene, that they will eventually develop cancer. The medical components of the counseling process for familial cancer risk are designed to address these concerns and include evaluation, recommendations, and treatment.

At the postoperative visit, the office nurse follows up on the information obtained at the initial meeting. Because of the possibility that WR may be at increased risk for a hereditary form of colon cancer based on his family history, the nurse discusses with the physician referral to the cancer-genetics specialist. Both WR and his wife feel this is a good idea. WR says that he feels lucky that the cancer was "caught so early." He expresses a sincere interest in knowing whether he is at higher risk for developing other cancers as well, and wonders what he should tell his son, who is turning 18 years old. WR believes that, armed with adequate information, his son might be vigilant about attending to symptoms and might increase his health surveillance (e.g., by undergoing fecal occult blood tests and colonoscopy at an earlier age than he would have otherwise).

The cancer-genetics nurse specialist in the clinic works with WR to obtain even more in-depth documentation of the family's experience with cancer. The nurse specialist and the genetic counselor meet with WR and his wife to provide education and counseling about hereditary colon cancer (Giardello, 1997).

FAMILIAL CANCER-RISK COUNSELING

Definition

Familial cancer-risk education and counseling is a communication process between health care professionals and patients concerning the occurrence, or risk of occurrence, of cancer in the patients' families (Peters, 1994a). Family members who participate in cancer-risk counseling are often seeking more information about the hereditary basis of cancer in their family and about available genetic testing and advice regarding medical surveillance. The current philosophy is that these needs can best be met by a multidisciplinary team consisting of professionals with expertise in oncology, surgery, genetics, nursing, and counseling psychology (Biesecker et al., 1993).

Medical Evaluation

Obtaining an accurate medical and family history from patients at risk for hereditary cancer is crucial to making the diagnosis of a hereditary syndrome. Not uncommonly, family members at risk for cancer already exhibit symptoms of malignancy at the time of genetic-risk assessment.

Medical Recommendations

For patients identified as having an increased cancer risk, the physician can customize surveillance and early-detection recommendations, taking into account a patient's particular medical and family history. The genetic counselor or genetic nurse specialist might discuss the medical recommendations for surveillance and treatment in lay terms and can help the family to prioritize and implement these recommendations. These specialists also contribute to developing a patient plan of care customized to individual needs, a plan that includes specific health promotion and disease-prevention guidelines. The document "Guidelines for Early Detection of Colorectal Polyps and Cancer" is one example of established guidelines for cancer prevention and health promotion (Byers et al., 1997). Additionally, the counselor has an opportunity to address patient anxieties and to explore obstacles to surveillance or treatment (Peters, 1994b).

Medical Treatment

Traditional medical treatments of cancer consist of surgery, chemotherapy, and radiation therapy. At this time, gene therapies for cancer have been applied in research settings only, for purposes of diagnosis, prognosis, and monitoring for residual disease. A patient's knowledge of his or her genetic-risk status may influence the treatment decision-making process. The nurse will have the opportunity to address patient concerns and anxieties, to discuss coping and support mechanisms, and to facilitate incorporation of new genetic information into the patient's family life.

The information obtained and discussed indicates that WR meets the criteria (presented in Table 8.2) for being a member of a family with hereditary nonpolyposis colon cancer (HNPCC). Not every member in such families inherits the genetic susceptibility for developing cancer, but in families with a documented history of HNPCC, each person who has a parent with HNPCC has a 50 percent chance of inheriting from that parent the gene for HNPCC.

TABLE 8.2

Amsterdam Criteria for HNPCC*

Pattern of CRC[†] indicative of HNPCC includes:

- Three or more relatives with histologically verified CRC through at least two successive generations

- One person a first-degree relative of the other two

- In at least one affected person, diagnosis made before age 50

*Hereditary nonpolyposis colon cancer
[†] Colorectal cancer

Source: Modified from H. Vasen, J. Mecklin, P. Meerakhan, et al., The international collaborative group on hereditary nonpolyposis colorectal cancer. Dis. Colon Rectum 34:424–428, 1991.

WR asks to learn more about the possibility of having genetic testing done to determine whether he has inherited the gene for HNPCC. He explains that he read in the newspaper about the availability of HNPCC testing. WR learns that genetic testing for HNPCC is being offered in this clinic as part of a research protocol. Because WR met the criteria for study entry, the nurse provides WR and his wife with more information about the risks and benefits of the testing. He needs this information to make an informed decision about whether to pursue genetic testing for HNPCC. The nurse is aware, as she is providing the information, that each person is unique and that a psychosocial assessment of the patient is an important component of total patient care.

Psychosocial Assessment Issues

Familiarity with several areas of psychosocial assessment may aid the nurse in understanding and facilitating the patient's decision making about, and adjustment to, cancer-susceptibility testing. The nurse should evaluate the patient's response to cancer (if he or she is affected); his or her emotional responses to a family history of cancer; the patient's biological and emotional closeness to affected relatives; and the patient's beliefs and fears about cancer (Peters, 1995).

The nurse can explore the meaning of cancer to the individual and to the family. For example, some people may consider cancer a punishment. For others, a high-risk status is sometimes assumed, based on physical or psychological resemblance to an affected individual. It has been observed that information offered by geneticists that is not in accordance with family explanations and expectations may be disregarded (Grosfield et al., 1996; Kenen, 1980). By eliciting from patients underlying beliefs about the causes of cancer prior to educating them about the genetic contribution to cancer, the nurse can help to foster better integration of new health information into existing health beliefs and behaviors.

Assessment also includes analysis of a patient's response to cancer and perception of risk. Individual responses that have been observed include anxiety, anger, fear of developing cancer, fear of disfigurement and dying, grief, lack of control, negative body image, comparisons to affected relatives, fear of altered sexual functioning, and a sense of isolation (Baker, 1991; Kelly, 1992; Lynch et al., 1994;

Lynch & Watson, 1992; Peters, 1995). Feelings of grief over loss of a significant loved one to cancer may reemerge during stressful times, such as during pregnancy, at the time of a medical appointment, or on the anniversary of a loss. Anxiety and fear, in particular, can have a major impact on daily activities, life decisions, and health care behavior (Lerman et al., 1993). Other psychological outcomes observed include competence or resilience, which may contribute to self-esteem and positive coping styles (American Society of Clinical Oncology, 1996; Lerman et al., 1991). Further research is needed to explore how individuals and families at increased risk incorporate and respond to knowledge of their risk status.

Genetic disorders may affect relationships among family members. Various members may react differently to genetic-testing information, depending on whether the family has known for years that it has a hereditary cancer predisposition or is receiving such information for the first time (Bert, 1996). Nursing assessments of marital and family relationships can help the nurse to anticipate interpersonal problems and to identify people who might need additional psychosocial support.

WR and his wife believe that it is important for them to learn whether WR has a gene mutation commonly found in families with HNPCC. The most common genes are MSH2 and MLH1. WR tells the nurse that they are interested in testing mainly because of their son. They want him to know whether he has inherited the susceptibility gene for HNPCC from WR, so that if he has, he will be able to participate early in surveillance procedures. If their son has not inherited the gene for HNPCC, then WR and his wife would be relieved of worry, as would their son. The nurse explains to WR and his wife that even if a gene mutation is found in WR, the test cannot be offered to his son until he is 18 years old. Genetic testing of adolescents for adult-onset illnesses is not offered (ASHG, 1995). WR expresses disappointment that genetic testing cannot yet be offered to his son but states that he is glad that testing is a health option that his son can consider in the future.

Both WR and his wife have questions about the impact of test results on their health-insurance coverage. They say that they are aware that in their state no legislation has yet been passed that addresses the use of genetic information in coverage decisions made by their insurance company. This concern has caused them to reflect carefully on their decision to undergo genetic testing at this time.

After the educational and counseling sessions with WR and his wife have been completed, the nurse begins the communication and collaboration process with WR's primary health care provider. A plan of care that addresses both the physical and psychological needs of WR and his family will be developed in collaboration with the primary care provider.

Plan of Care

The nurse can provide follow-up education to discuss the medical recommendations, to help the family to prioritize and implement the recommendations, to coordinate medical appointments, and to facilitate referrals. The plan of care incorporates attention to the need for counseling interventions related to individual and family responses to cancer risk or development. Coordination of care often involves several disciplines and multiple procedures. The nurse is aware that the counseling and testing process heightens emotions, and therefore sensitivity and attention to communication are essential elements in facilitating a quality plan of care.

Psychosocial Interventions

A variety of interventions that are commonly used during the genetic-counseling process may be helpful during familial cancer-counseling sessions. Nurses will participate in this counseling process.

Grief counseling is essential. Because present emotions are influenced by past experiences, adequate counseling time is allowed to elicit historical information. Discussion of family history often provides an opportunity for patients to remember and to grieve for their personal losses.

Supporting a realistic sense of hope is another important counseling goal. Many patients who enter the risk-counseling process may feel fatalistic about developng and dying from cancer. Providing supportive explanations that demonstrate the probabilities for developing or not developing cancer can be very helpful. This information may motivate people to engage more consistently in health-promoting behaviors.

Dispelling misconceptions is an additional intervention that can help to relieve unnecessary emotional burdens. For example, some women worry that a bruise may mean that they will develop breast cancer. For those women, providing information about the biological basis for cancer can be of great psychological benefit. Other interventions include relaxation training and biofeedback to reduce anxiety.

Patient support groups are a psychosocial intervention that has been successful in fostering adjustment both to rare genetic diseases and to cancer. Studies have found that support-group experience improves surveillance in high-risk patients and that these behavioral effects persist for several years. It has been observed, for example, that women with metastatic breast cancer and patients with malignant melanoma do better both medically and psychologically when they attend a support group with a trained psychologist. This group-therapy model has been expanded to include women who have an increased risk of developing breast cancer (Peters, 1994a, 1994b).

WR decides to give a blood sample for genetic testing. He says that he does not feel in need of additional support beyond his family at this time but is reassured to know that other resources for information about hereditary colon cancer exist on the internet and in pamphlets. He decides to let his family physician know of his decision to enter the clinical trial. The nurse reminds WR that he can choose the people whom he wants to inform of his participation in the research trial and assures him that confidentiality is an important aspect of all medical records and documentation related to genetic testing, because misuse of such information can occur.

Once the session is completed, the cancer-genetics team takes the opportunity to review the outcome of the session. They review the ways in which their clinic resources were used in WR's situation and the effectiveness of the information provided.

GENETIC SUSCEPTIBILITY TESTING

Testing for inherited cancer susceptibility, like other genetic testing, poses the challenges of profound personal and family implications and the possibilities of stigmatization, discrimination, misunderstanding, and misuse of information. Until recently, genetic susceptibility testing was done in the context of research studies in which the indications for the testing were based on the questions that the research studies proposed to answer. Patients considered to be the most appropriate candidates for cancer susceptibility testing are members of families with hereditary cancer syndromes for which the specific gene mutation has been identified.

A number of policy statements developed by genetics organizations urge that cancer-susceptibility testing be carried out with caution until more information is gained about the test characteristics and implications (American Society of Human Genetics, 1994; Biesecker et al., 1993; Evans et al., 1992, 1994; King et al., 1993; Li et al., 1992; Lynch & Watson, 1992; National Advisory Council for Human Genome Research, 1994). Despite these recommendations, cancer susceptibility testing is being offered commercially for common cancers and other disorders in larger populations and within primary care practices.

Genetic Counseling for Susceptibility Testing

The current approach to cancer susceptibility testing places a strong emphasis on the importance of pretest counseling, the multidisciplinary team approach, and the necessity for follow-up for tested family members (Biesecker et al., 1993). At present, testing is being offered mainly to self-referred persons within high-risk families or in high-risk registries. The principles guiding cancer susceptibility testing have been described by Schneider (1994) and are important for nursing consideration.

Pretest Counseling for Cancer-Susceptibility Testing

Pretest counseling and the informed-consent process are important elements of cancer-susceptibility testing and provide the foundation for notification of test results. The best ways to educate and counsel individuals considering genetic-susceptibility testing for cancer have not yet been established. The following components, as outlined by Schneider (1994), offer one framework for counseling:

- exploration of motivation for and expectations from testing
- assurance of informed consent during pretest counseling

- review of basic medical and genetic facts
- discussion of possible results and implications (accuracy and limitations of test results, risks and benefits of learning results, extent and limitations of privacy and confidentiality)
- assessment of coping resources and formulation of coping strategies
- discussion of need for follow-up counseling and evaluation

Motivation and Expectations

People's motivation for seeking cancer-susceptibility testing varies. Some value knowledge both for its own sake and for a sense of control over life that this knowledge implies. Others may hope to put uncertainty to rest. Some might wish to reduce or eliminate expensive and risky medical surveillance programs, which they could undertake if they were to test negatively. Many undergo testing to assist with research endeavors or to learn about risks for their children.

Motivation for testing is sometimes based on misconceptions about the nature of testing (e.g., a belief that the test will provide a definitive answer about cancer risk). Some may believe, for example, that the medical community would not offer cancer susceptibility testing if there were nothing that one could do to alter one's risk. These test expectations might be a source of disappointment later if the testing fails to fulfill a person's underlying needs and desires. Therefore, assessing participants' motivations for testing is important, so that subsequent information can be tailored to their needs and expectations.

Informed Consent

One of the most crucial components of genetic testing for cancer susceptibility is that of informed consent. Informed consent (as described in Chapters 4 and 9) refers to a communication process (usually both verbal and written) between health care providers and patients in which the patients are provided with sufficient information to decide whether to be tested and, if so, when and how. Throughout the process, each patient is assured of protection of privacy and voluntary participation without coercion from others. During counseling, professional and family values are taken into account. Cognizant of these values, the counselor makes all attempts to present a balanced view of the risks, benefits, and limitations of genetic-susceptibility testing. Families may have cultural beliefs about health and illness, testing biases, and spiritual beliefs that might influence their choices. These also should be explored and discussed.

Provision of Basic Medical and Genetic Facts

Prior to testing, all patients are given information about the basic laws of inheritance; genetic heterogeneity (more than one gene causing the same condition); gene penetrance (not everyone with a mutation will manifest cancer); variable phenotype expression (different people with the same mutation will exhibit different sequelae); and the occurrence of nonhereditary cases even within families with a hereditary condition (Richards et al., 1995). The counselor also explains that having a gene mutation for a particular cancer does not mean that a person actually has cancer; rather, it means that the individual has a higher chance of, or a *predisposition* to, develop cancer within his or her lifetime, as compared to other people who do not carry the gene. Medical options that are available for risk reduction, early detection, and treatment are also presented.

Susceptibility-Test Implications

During the course of pretest counseling, participants are prepared equally for negative, positive, or indeterminate results of genetic-susceptibility testing. Each of these results carries with it associated medical, psychosocial, and genetic implications (abstracted in Table 8.3). Implications differ, depending on several factors; for example, the implications for a patient who is being tested and who already has received a diagnosis of cancer would be different from those for a patient who has been successfully treated for cancer and who is choosing testing to confirm his or her risk and to learn whether there is also an increased risk for other medical problems. Whether a cancer-causing mutation has already been identified in a family or whether the presenting patient is the first family member to be tested also significantly affects test implications. Test participants may have greater confidence that negative results are truly negative when a mutation is already known to occur in a given family than when a mutation must be discovered and shown to cause cancer in the family.

TABLE 8.3

Medical and Psychosocial Implications of Cancer Genetic Testing

POSITIVE RESULT	NEGATIVE RESULT

Medical

Benefits

Cancer detection and prevention options

Risks

Knowledge of increased cancer risk

Delay in seeking medical treatment for cancer

Psychosocial

Benefits

Reduction of uncertainty

Increased support

Eligibility of children for DNA testing

Possibility of offering cancer surveillance to children

Increased ability to plan for the future

Risks

Anxiety

Depression or suicide

Lowered self-esteem

Lowered life goals

Strained relationships with family (spouse, children)

Loss of insurance

Discrimination by employers, colleges, social service

Costs of cancer surveillance

Medical

Benefits

Extra surveillance not needed, because cancer risk same as for the average person

Risks

Delay in seeking care resulting from misunderstanding of cancer risk

Psychosocial

Benefits

Feelings of relief and joy that children cannot inherit the gene

Financial savings from less cancer surveillance

Feelings of increased ability to plan for the future

Risks

Survivor guilt

Depression because cancer risks are not the cause of problems

Strained relationships with family members

Source: Adapted from K.A. Schneider, Counseling About Cancer: Strategies for Genetic Counselors. *Wallingford, PA: National Society of Genetic Counselors, 1994.*

Notification of Risk

Explaining to patients their chances for inheriting or developing cancer is generally reserved for a specialized risk-assessment clinic or a clinic staffed with professionals trained to perform risk analysis. Once a patient's risk for inheriting or developing cancer is estimated, the patient is given information in a clear manner. Several factors may interfere with a patient's ability to comprehend the genetic risk information:

- difficulty in interpreting abstract laws of probability (the percent chance of inheriting or developing cancer),
- preconceived beliefs about the actual risk, and
- a tendency to simplify information in order to resolve uncertainties about risk.

For these reasons, health care personnel must continally assess the patient's understanding of information that is presented to help them to determine the patient's perception of risk and to clarify any misperceptions.

Information about the risk of inheriting or developing cancer is communicated by both verbal and visual means. Percentages and fractions are offered, with a focus on what risk means in terms of the patient's chances of inheriting a cancer predisposition versus the chances of developing cancer. Some reported patients' emotional responses to receiving risk information are relief, anxiety, depression, fear, confusion, grief, loss, anger, shame, guilt, and worry about other family members. Positive test results may alter self-perception and life goals and interfere with one's ability to seek appropriate medical care (Biesecker et al., 1993; Grosfeld et al., 1996; Lynch & Watson, 1992).

Follow-Up Counseling

Nurses have a major role in supporting test participants during and after carrier susceptibility testing, so referral to appropriate specialists and support groups is an important role for them. No specific variables have yet been found to predict which patients will adapt more readily to new genetic information or which ones may experience long-term difficulty in adjusting to test results. Through the

National Institutes of Health Ethical, Legal, and Social Issues Hereditary Consortium, a number of studies are being conducted to assess such variables (USDHHS, 1996).

WR's genetic test results come back from the laboratory several months later. The nurse calls WR and his wife and invites them to come to the clinic to discuss the test results. When WR asks whether the nurse can provide this information over the phone, the nurse responds that meeting in person with patients to discuss test results is a part of that clinic's protocol and that this policy allows the best opportunity to review test results, ask and answer questions, and clarify concerns. WR and his wife come to the clinic and are told that WR does have a gene mutation that predisposes him to develop HNPCC. WR and his wife indicate that they are relieved to have this information and begin talking with the nurse about how they will now share this information with their son. WR says he does not believe that having these results will modify anything for him personally, as he was already participating in frequent colon-surveillance procedures.

WR tells the nurse that he is contemplating sharing this information with other family members, such as his brother, who might want to find out about their risk for inheriting the gene for HNPCC. He explains that although his brother does not yet have any symptoms, WR feels that he needs to share this information with him so that his brother can determine whether he wants to learn more about genetic testing and his options.

The nurse calls WR 2 weeks later to follow up on their visit and to inquire whether WR and his wife have any remaining questions or concerns. WR says that he is somewhat depressed, because he knows that he may have passed the gene for HNPCC on to his son. Nonetheless, he is generally satisfied that he decided to be tested, because he now has specific information that his son can consider. WR explains to the nurse that he has some suggestions regarding how the information was presented in the clinic and is putting these in writing so that the team can use his suggestions for other patients and their families.

Evaluation

Evaluation of the effectiveness of all components of nursing participation in educating and counseling patients and families at risk for hereditary cancer susceptibility is essential. In particular, research in skill development, educational design for patients and staff, and intervention effectiveness will be helpful. Prospective studies involving long-term follow-up should be considered in evaluating each of these components. The impact of each of these factors must be evaluated with an eye toward continuous improvement in patient care. Quality clinical and psychosocial outcomes must be determined and then measured to ensure successful application of new genetic technologies.

Resources

Accessible resources through which nurses can learn more about cancer genetics are being developed. The International Society of Nurses in Genetics and the Oncology Nursing Society are two professional societies that offer guidelines, educational support, and expert consultation regarding the development of this new role for cancer nurses. Appendix C offers additional resources.

CONCLUSION

Appreciating the basics of cancer genetics and the components of cancer-risk assessment and susceptibility counseling and testing can have a variety of benefits for patients, families, and nurses, among them the following:

- Gaining a conceptual appreciation of the ways in which various genetic changes in cancer fit into a cohesive pattern of cancer susceptibility.
- Becoming familiar with new genetic technologies as they are (or will be) incorporated into clinical oncology practice.
- Gaining a better understanding of the biological mechanisms underlying cancer prevention, diagnosis, treatment, and surveillance to ultimately decrease cancer morbidity and mortality.

- Stimulating thinking about ways of educating patients about these changes.
- Improving the capability of accurate risk assessment.
- Meeting the medical, emotional, and genetic needs of individuals from families with genetic predisposition to developing cancer.
- Informing families about research protocols for cancer-risk assessment and registries available for consideration.
- Participating in strategies in which designer treatments will be modified for the individual, depending on the genetic changes encountered in the tumor and on the person's underlying genetic constitution.

Scientific discoveries are enabling scientists to increase their understanding of the genetic aspects of cancer. Genetic testing offers an immediate means by which to apply genetic science to clinical practice involving patients and families who are at increased risk for inherited cancer susceptibilities. Nurses who participate in educating and counseling patients about these discoveries will want to learn of such advances in the field of oncological genetic research. In time, the scope of practice and competencies of the basic and advanced oncology nurse will be developed to guide nurses in other care settings as they begin to integrate genetic knowledge into their practices.

SUMMARY POINTS

Assessment
- patient identification and screening
- assessment of risk factors
- risk analysis
- diagnosis
- psychosocial assessment

Counseling
- a communication process between health care professionals and patients concerning the occurrence, or risk of occurrence, of cancer in families

- entails discussion of medical evaluation, recommendations, and treatment
- involves presentation of and follow-up to a plan of care, including psychosocial intervention, genetic intervention, and referral

Elements of the genetic testing process
- pretest counseling, risk notification, and follow-up counseling
- assurance of informed consent
- review of basic medical and genetic facts
- discussion of possible results and implications
- evaluation of effectiveness of counseling
- offer of additional resources

QUESTIONS FOR CRITICAL THINKING

1. Discuss the elements of cancer-risk assessment and susceptibility testing that you would share with a patient during pretest counseling.

2. How do you feel about the option of genetic testing for breast cancer in high-risk populations? for screening of the general population?

3. What other components of the nursing role should be addressed in a model of nursing care that integrates genetic knowledge into cancer care?

REFERENCES

American Society of Clinical Oncology (1996). Statement of the American Society of Clinical Oncology: genetic testing for cancer susceptibility. *J. Clin. Oncol.* 14:1730–1736.

American Society of Human Genetics (ASHG) (1994). Statement of the ASHG on genetic testing for breast and ovarian cancer disposition. *Am. J. Hum. Genet.* 55:1–4.

American Society of Human Genetics Board of Directors and the American College of Medical Genetics Board of Directors Consensus Committee (1995). Points to consider: ethical, legal and psychosocial implications of genetic testing in children and adolescents. *Am. J. Hum. Genet.* 57(5):1233–1241.

Baker, N. C. (1991). *Relative risk: living with a family history of cancer.* New York: Penguin Books.

Bert, T. (1996). Review of article on genetic counseling for familial adenomatous polyposis. *Oncology* 10(1):97–98.

Biesecker, B. B. (1997). Psychological issues in cancer genetics. *Semin. Oncol. Nurs.* 13(2): 129–134.

Biesecker, B. B., Bohenke, M., Calzone, K., et al. (1993). Genetic counseling for families with inherited susceptibility to breast and ovarian cancer. *J.A.M.A.* 269(15):1970–1974.

Byers, T., Levin, B., Rothernberger, D., et al. (1997). *Amercian Cancer Society guidelines for screening and surveillance.*

Calzone, K. (1997). Predisposition testing for breast and ovarian cancer susceptibility. *Semin. Oncol. Nurs.* 13(2):82–90.

Claus, E. B., Risch, N., Thompson, W. D. (1994). Autosomal dominant inheritance of early onset breast cancer. *Cancer* 73:643–651.

Croyle, R., and Lerman, C. (1993). Interest in genetic testing for colon cancer susceptibility: cognitive and emotional correlates. *Prev. Med.* (22): 284–292.

Evans, D. G. R., Fentiman, I. S., McPherson, K., et al. (1994) Familial breast cancer. *Br. Med. J.* 308: 183–187.

Evans, D. G. R., Ribiero, G., Warrell, D., and Donnai, D. (1992). Ovarian cancer family and prophylactic choices. *J. Med. Genet.* 29:416–418.

Fraser, M., Calzone, K., and Goldstein, A. (1997). Familial cancers: new and evolving clinical issues. *Oncol. Nurs. Updates* 14(3), Hubbard, S. M., Goodman, M., and Knobf, M., eds. Cedar Knolls, NJ: Lippincott-Raven.

Gail, M. H., Brinton, L. A., Byar, D. P., et al. (1989). Projecting individualized probabilities of developing breast cancer for white females who are examined annually. *J. Natl. Cancer Inst.* 81:1879–1886.

Geller, G., et al. (1977). Genetic testing for susceptibility to adult-onset cancer. The process and content of informed consent. *J.A.M.A.* 277(18):1467–1477.

Giardello, F. (1997). Genetic testing in hereditary colorectal cancer. *J.A.M.A.* 278(15):1278–1281.

Green, M. (1997). Genetics of breast cancer. *Mayo. Clin. Proc.* 72:54–65.

Grosfeld, F. J. M., Lips, C. J. M., and TenKroode, H. F. J. (1996). Psychosocial consequences of DNA analysis for MEN type 2. *Oncology* 10(2): 141–146.

Kelly, P. T. (1992). Breast cancer risk analysis: a genetic epidemiology service for families, *J. Genet. Counsel.* 1(2):155–168.

Kenen, R. H. (1980). Negotiations, superstitions and the plight of individuals born with severe birth defects. *Soc. Sci. Med.* 14A:279–286.

King, M. C., Rowell, S., and Love, S. M. (1993). Inherited breast and ovarian cancer: what are the risks, what are the choices. *Am. J. Med. Genet.* 269:1975–1980.

Lerman, C., Rimer, B. K., and Engstrom, P. F. (1991). Cancer risk notification: psychological and ethical implications. *J. Clin. Oncol.* 9(7):1275–1282.

Li, F., Garber, J., Friend, S., et al. (1992). Recommendations on predictive testing for germline p53 mutations among cancer-prone individuals. *J. Natl. Cancer Inst.* 84(15): 1156–1160.

Lynch, H. T., Lynch, J. F., Conway, T., and Severin, M. (1994). Psychological aspects of monitoring high risk women for breast cancer. *Cancer* 74:1184–1192.

Lynch, H. T., and Watson, P. (1992). Genetic counseling and hereditary breast/ovarian cancer. *Lancet* 229:1181.

National Advisory Council for Human Genome Research (1994). Statement on use of DNA testing for presymptomatic identification of cancer risk. *J.A.M.A.* 271(10):785.

Peters, J. A. (1994a). Familial cancer risk, part 1: Impact on today's oncology practice. *J. Oncol. Mgt.* 3(5):20–30.

Peters, J. A. (1994b). Familial cancer risk, part 2: Breast cancer risk counseling and genetic susceptibility testing. *J. Oncol. Mgt.* 3(6):14–22.

Peters, J. A. (1995). Breast cancer genetics: relevance to oncology practice. *Cancer Control* May/June:195–207.

Richards, M. P. M., Hallowell, N., Green, J. M., et al. (1995). Counseling families with hereditary breast and ovarian cancer: a psychosocial perspective. *J. Genet. Counsel.* 4(3):219–232.

Schneider, K. A. (1994). *Counseling about cancer: strategies for genetic counselors.* Wallingford, PA: National Society of Genetic Counselors.

U. S. Department of Health and Human Services (USDHHS), National Institutes of Health, National Center for Human Genome Research (1996). *Review of the Ethical, Legal and Social Implications Program 1990–1995.* Washington, DC: USDHHS.

Vasen, H., Mecklin, J., Meerakhan, P., et al. (1991). The international collaborative group on hereditary nonpolyposis colorectal cancer. *Dis. Colon Rectum* 34:424–428.

CASE STUDIES

The case studies* presented here describe patients and families undergoing predisposition testing for breast cancer. Currently, predisposition testing for breast cancer usually occurs in a research setting and is offered by a team of specialists, including nurses. Nurses providing primary care may encounter patients who have questions about this testing, who are seeking referral, or who are already undergoing predisposition testing. The case studies allow consideration of a variety of nursing activities in each of these situations.

Inherited susceptiblity to breast cancer has been recognized in the medical literature since 1866; however, the specific predisposing genes were not identified until recently. Cancer susceptibility genes are currently believed to be the cause of 5–10 percent of all breast and ovarian cancers. BRCA1 and BRCA2 are the best-recognized mutations. BRCA1 gene mutations, for example, have been found in approximately 45 percent of all families who have inherited breast cancer and in 90 percent of families in which both breast and ovarian cancer are present. Both men and women may carry breast-cancer gene mutations. Men who have a BRCA1 gene mutation do not have an increased risk for developing breast cancer, whereas men who have the BRCA2 gene mutation are at increased risk. In either situation, children of such men would be at increased risk for inheriting the predisposing gene. It is also known that 1 percent of the Ashkenazi Jewish population has a BRCA1 gene mutation.

Assessment and analysis of family history and ethnic background are essential for identifying women and families who may have, or who are at increased risk for developing, an inherited breast-cancer syndrome. Equally important are identifying women and families who are not at increased risk and providing appropriate information about cancer risk and surveillance. Predisposition testing is available and can be offered to members of families identified as having an

---

Case studies for this chapter were adapted with permission from cases presented by Kathleen Calzone, RN, MSN, at the Oncology Nurses Fall Institute, Creating an Oasis, during her presentation "Cancer Genetics for Nursing: Creating a Map to the Future."

increased risk of developing breast or ovarian cancer. Testing is not appropriate for everyone, and population screening is currently not recommended.

CASE STUDY 8.1

DISCLOSURE OF RESEARCH FINDINGS

Ms. T, a 32-year-old woman, arrives for an annual primary care visit and reviews her recent medical and family history with the nurse. Ms. T informs the nurse that she is the only female in her nuclear family who has not developed breast cancer. She relates that she has two sisters, one of whom has already died (at age 34) from breast cancer; the other, age 37, is coming into the hospital clinic today for chemotherapy.* Ms. T reports that she has decided to undergo bilateral prophylactic

*A family history of breast cancer has been recognized as a risk factor since the 1800s. Advances in molecular genetics have provided evidence that cancer-susceptibility genes may be responsible for 5–10 percent of all breast and ovarian cancers (Green, 1997). The BRCA1 gene for breast cancer, located on chromosome number 17, and the BRCA2 gene, located on chromosome number 13, are two genes known to predispose patients to breast, ovarian, and other cancers. Predisposition testing is now available for breast cancer and can be offered to patients and families who are at increased risk because of family history. Identification and counseling of these families are emerging nursing-practice activities (Calzone, 1997; Geller et al., 1997).

mastectomies and wants to discuss this issue with her physician. She recalls that she and her sister underwent many tests during the course of her sister's illness, and Ms. T wonders whether any of these tests would provide information about her chances of developing breast cancer.

The nurse later learns that Ms. T, her sisters, and family participated in breast-cancer research studies and that genetic studies were done. The research team has the results of the research studies, which could help to clarify Ms. T's status, but there was never any intention to provide results to the family and, in fact, that had been specifically stated in the informed-consent form.

- What is your responsibility to Ms. T at this point?
- How does the fact that a result is already available influence the situation for Ms. T?

Discussion

The nurse has a responsibility to pursue an answer to Ms. T's question, and to serve as an advocate for her care. The nurse can discuss Ms. T's concerns with the primary care physician, and together, they can explore with the research team the following issues:

- the research team's obligation to reveal that test results are available
- unique counseling considerations such as research protocol, confidentiality, and the fact that test results are already available
- consideration, in caring for Ms. T, of a known result that might be helpful to the patient

One possible outcome is that, after discussion, the research team decides to reveal that test results are available that could further define Ms. T's status with regard to breast cancer. On the basis of family information, they offer Ms. T counseling and presymptomatic testing; she decides to proceed with testing, and the results are disclosed to her. She is found not to have inherited the familial gene for breast cancer and so decides to cancel her prophylactic surgery.

The nurse in this setting might coordinate testing and counseling in collaboration with the research and counseling teams, support Mrs. T in her decision-making process, and ensure that the patient has adequate follow-up support, because some women whose test results are negative may experience feelings of guilt when other family members are affected.

THE RIGHT TO REFUSE TO UNDERGO TESTING

Mr. N has a family of four daughters. He also has three sisters and a brother. Two of his three sisters were diagnosed with breast cancer in their forties, chose to have BRCA1 breast cancer testing, and were found to carry the BRCA1 gene. Mr. N's daughters learn of this history and want to be tested to find out their status. Mr. N does not want to be tested himself.* He tells the nurse at the clinic, "I have a right not to be tested, and I don't want to know! I care about my daughters, but I am afraid to find out for them." One of the daughters calls the nurse and asks for help.

*The BRCA1 breast cancer gene is associated with an increased lifetime risk of breast cancer in women who carry the gene mutation. Carriers of this gene may also develop ovarian cancer. There also appears to be a small but increased risk for prostate cancer in men and colon cancer in men and women. Although a male may carry the BRCA1 gene, male breast cancer has not been associated with mutations in BRCA1 (Calzone, 1997).

- How might you respond to Mr. N?
- What could you say to the daughter who called?
- What are the counseling issues for this daughter that are unique to this situation?

Discussion

The nurse has an obligation to support Mr. N in his right not to know; however, the nurse can further explore with Mr. N his concerns and fears for his daughters in an effort to help this patient to clarify his feelings. The nurse also refers Mr. N and his family to the cancer-genetics clinic for further information and counseling.

When the daughter calls the nurse, the nurse can directly refer the daughter and her sisters to the cancer-genetics clinic for further information and counseling.

Possible outcomes are:

- All four daughters choose to be tested and to learn their results and are informed that they did not inherit the BRCA1 gene alteration associated with cancer in the family.
- All four daughters choose to be tested and, though three learn that they did not inherit the BRCA1 gene mutation, one daughter learns that she did. This information has implications for her management and treatment. It also provides information about her father's status: he is a carrier of the BRCA1 gene. In this scenario, the nurse refers the daughter back to the genetics clinic to discuss further the ways that she can incorporate this new information in her life to benefit her health and to discuss with the counseling team ways that she might talk with her father about the test results.

In either of these settings, the nurse could participate in supportive and follow-up counseling, coordinate with the genetics clinic the daughter's and Mr. N's care, and ensure continuity of their care.

CASE STUDY 8.3

THE DUTY TO NOTIFY OTHER FAMILY MEMBERS

Mrs. F has a family history of breast cancer; two sisters died in their thirties, and a maternal aunt died in her twenties from this disease. Mrs. F is offered BRCA1 testing and wants to speak with her primary care physician about pursuing testing. During the intake interview, she informs the nurse that when she was young and unmarried, she had an unplanned pregnancy. Her baby girl was placed for adoption. Only one of Mrs. F's sisters knew of this pregnancy, because she moved out of town to live with relatives throughout the pregnancy.

Mrs. F is currently married and has children. She tells the nurse that her husband and her other children do not know of this baby. Mrs. F says she feels a strong obligation to the child, whom she has never seen since giving her up for adoption, and would want her to have the information about her potential for carrying the BRCA1 gene if she (Mrs. F) is found to have the gene mutation. She states that she does not want the adopted daughter to try to contact her as an outcome of this information. Rather, she wants to ask the cancer-counseling team if they will serve as the intermediary between her and the adoption agency in an effort to protect her confidentiality.

- What is your role in this situation?
- What is Mrs. F's duty to inform her daughter?

Discussion

The nurse's role in this situation is to support Mrs. F in her efforts to inform her daughter about a positive BRCA1 test result. The nurse offers to talk with Mrs. F's physician about referring her to the cancer-genetics counseling team for further discussion about BRCA1 testing and about the ethical issues surrounding Ms. F's desire to inform her daughter and her wish not to be contacted. After the visit (with permission), the nurse makes a referral to the cancer-genetics clinic and informs the counselor of Mrs. F's issues and concerns.

The nurse learns that Mrs. F proceeds with testing and that she does not have the BRCA1 gene mutation associated with cancer in her family. The nurse also learns that the cancer-genetics counseling team had decided that if the test results showed that Mrs. F was a carrier of the BRCA1 gene, the case would have been presented to the ethical advisory board of the health care facility for further discussion about the duty to disclose the information to Mrs. F's daughter and about Mrs. F's desire not to be contacted.

CASE STUDY 8.4

GENETIC TESTING FOR CANCER IN LIGHT OF PREVIOUS DECISIONS

Ms. H, age 43, has had bilateral prophylactic mastectomies because of a family history of breast cancer in three of her father's sisters and ovarian cancer in two other paternal aunts and two of their daughters. She comes for a routine office visit and informs the nurse that she has been reading about predisposition testing for breast and ovarian cancer* and would like to consider this testing because of her increased risk for ovarian cancer.

Predisposition testing for BRCA1, a gene for breast cancer also associated with an increased risk for ovarian cancer in those who carry it, is available commercially, but there is no consensus in the medical and genetic community as to its use outside of clinical research trials. Predisposition testing has both risks and benefits, and therefore informed consent is the essential component of all genetic testing. As in other areas of nursing practice, the nurse assumes the role of advocate to ensure that the patient is given appropriate, understandable, and complete information with which to make an informed health decision (Biesecker, 1997; Calzone, 1997).

- What is your role in counseling a woman in this situation?
- What strategies could be used to help her in her decision making?

Discussion

The nurse informs Ms. H that testing may be helpful to her in this situation; however, more information about her family history and whether predisposition testing has been done on any other family members will be needed. The nurse explains that patients who have an indication for, or interest in, predisposition testing for breast cancer are usually referred to the cancer-genetics counseling team for further information and assistance. The counseling team will provide genetic evaluation and supportive counseling before, during, and after the testing process. The nurse also discusses information about the patient's risk for ovarian cancer and treatment options.

Ethics, Genetics, and Nursing Practice

Christine Grady
PhD, RN, FAAN

. . . These intrepid discoverers . . . are scientific Magellans still at the beginning of a voyage that will map the whole human genome. The profound hope is that knowing will lead to healing . . . But today we have an answer to a scientific puzzle that comes with a hundred new questions.

Ellen Goodman, Washington Post, *10/28/94, p. A21*
(on the occasion of the discovery of BRCA1)

GENE CARE LINK

For thought-provoking information about ethical and social implication of genetics information, go to http://www. jbpub.com/clinical-genetics.

OBJECTIVE

Identify ethical and societal issues and complexities related to genetic information and technology.

RATIONALE

Nurses will be involved in offering, discussing, recording, and reporting genetic information. Having an understanding of some of the important ethical and social issues related to these activities will help nurses to provide appropriate resources and support to patients, families, and communities.

APPLICATION ACTIVITIES

- Support confidentiality and privacy of genetic information.
- Ensure autonomy in patient decision making, including the right to decline genetic testing.
- Participate in discussing anticipated benefits and risks of genetic tests.
- Participate in and support appropriate genetic counseling for patients and families throughout the genetic testing process.

INTRODUCTION

T*he ultimate goals of genetic research are to increase our understanding of the human genome and of the specific genes that underlie human diseases and traits, so that we can better prevent, diagnose, and treat genetic diseases. Genetic research is characterized by excitement and promise but also by uncertainty and some anxiety. Through the work of the Human Genome Project (HGP), a detailed genetic and physical map of the entire human genome will be made widely available. Identification and mapping of human genes will enable researchers to decipher the function of each gene and to explain how defects result in physiological harm or disease. Once the functions and malfunctions of genes are understood, methods of preventing or treating the resultant disorder can be developed and tested. The pace of genetic discovery is rapid, with scientific knowledge about genes far outpacing deliberations or well-considered conclusions as to how and when to apply this knowledge to improve health.*

THE RELEVANCE OF ETHICS

As science tries to simplify and understand nature, including the sequence and functions of genes and genetic mutations and disorders, the study of ethics helps us to reflect on the hidden complexities and implications of such knowledge and to make rational decisions about whether, how, and when to apply what we learn. Science, like most human endeavor, is value laden and has ethical, social, legal, and political implications. New scientific knowledge and technology present new possibilities, choices that may not have existed previously, and these choices are often accompanied by conflict or uncertainty about the ethical (or proper) thing to do.

Decisions about the pursuit and application of scientific knowledge should be made with consideration of the ethical dimensions of such choices.

Ethical questions occur in various realms: Individual concerns are primary when one is deciding whether a person should have a genetic test for the presence of the gene for Huntington's disease. In the institutional realm are such decisions as whether and under what conditions an institution or organization should offer this test to its patients. At the societal level are decisions regarding whether private or public health insurance should pay for a genetic test or for screening of a given population group. Although current and compelling ethical questions related to genetics exist at every level, this chapter will focus primarily on ethical considerations in the individual realm, because the individual is the main concern of clinicians.

APPROACHES TO ETHICS

Ethics is a branch of philosophy that deals with values related to human conduct, the rightness or wrongness of certain actions, and the goodness or badness of the motives or ends of such actions (Beauchamp & Childress, 1994). Ethics invites us to use reasoned analytical and critical approaches and precedents to answer questions about how we *ought* to behave in given situations and *why*.

Theories and principles of ethics guide us in searching for and appraising solutions to ethical problems, as do legal and ethical precedents. Ethical theories organize and justify our analyses in different ways, depending on the nature of each theory. Theories define what it means to act morally and attempt to articulate and to justify principles that can both guide moral decision making and be used as standards for the evaluation of actions and policies.

Four principles are widely accepted as being central to biomedical ethics: respect for autonomy, nonmaleficence, beneficence, and justice (Beauchamp & Childress, 1994; see Table 9.1). Additional principles to which one might appeal and that might be important in clinical situations include fidelity, confidentiality, and respect for people. The Institute of Medicine Committee on Assessing Genetic

Risks (Andrews et al., 1994, p. 247) emphasizes four "important ethical and legal principles: autonomy, confidentiality, privacy, and equity" that assist in the ethical analysis of relevant questions related to genetic testing. Principles are considered general action guides, within which there is room for interpretation and judgment in specific cases. Faced with the complex life circumstances and dilemmas of individual patients and families, clinicians can find these principles abstract and difficult to apply, interpret, or rank. Some have argued that a principle-based approach to ethics is inadequate for the complex, human dimensions of many clinical bioethical dilemmas. Other (non-principle-based) approaches to the analysis of clinical ethical problems have been proposed (see, for example, Devettere, 1995; Gert et al., 1996; Jonsen & Toulmin, 1988). A "care" ethic, built on the understanding that individuals are unique, that relationships and their value are crucial in moral deliberations, and that emotions and character traits have a role in moral judgment (Noddings, 1984), adds an important dimension to ethical guidance for clinicians. Ethical analysis might involve asking multiple and

TABLE 9.1

Principles Useful in Ethical Analysis

Respect for autonomy: an obligation to respect, and not to interfere with, the self-determined choices and actions of autonomous individuals

Nonmaleficence: an obligation to never deliberately harm another

Beneficence: an obligation to maximize benefits and minimize risks and to promote the welfare of others

Justice (equity): an obligation to be fair in the distribution of social goods such as health care or in respecting people's rights or laws

Fidelity: an obligation to keep promises, contracts, and other agreements

Confidentiality: an obligation to protect and not to disclose personal information provided in confidence by another

Privacy: the right to control one's own body, thoughts, and actions

Respect for people: an obligation to respect the capacities and differences in human beings and to act accordingly

varied questions about the implications of each new genetic discovery and its potential applications (see Table 9.2).

TABLE 9.2

Useful Questions in Ethical Analysis

What are the potential benefits, and who will benefit?

What are the potential harms, and who will be harmed?

Who makes the decisions? Who *ought* to make the decisions?

What is the fairest thing to do?

How does one maintain respect for the integrity and dignity of the involved individuals?

What rights or responsibilities should be protected or maintained?

What are the unique needs of the involved individual(s)?

What relationships should be taken into account?

What virtues should be considered or encouraged?

The outcome of careful ethical reflection and analysis, regardless of the approach, does not preclude the possibility that reasonable people will disagree; in fact, disagreement is not uncommon. Nonetheless, there is considerable value in the process of reflection, dialogue, and analysis: "the process of debate and scrutiny . . . is likely to produce the kind of thoughtful judgement that is always more valuable than simplistic conclusions reached without the benefit of careful, sustained reflection and discourse" (Reamer, 1991, p. 13).

OVERVIEW OF ETHICAL ISSUES COMMONLY CONFRONTED IN GENETICS

A large number of the ethical issues that challenge us in the context of genetics are not new. Many of these issues have received a great deal of ethical attention, reflection, and debate, and although they are pertinent to genetics, they are not restricted to this field. Among these issues are privacy, confidentiality, discrimination, prenatal testing, abortion, access to health care, justice in health care, informed consent and patient decision making, and the conduct of clinical research. The considerable body of literature on some of these issues in other health care and biomedical research contexts contributes valuably to deliberations regarding genetics. Yet genetic discovery and its application to clinical practice present unique dimensions to some of these ethical questions that also require attention. For example, similar to any other personal medical information, medical information about genetic susceptibility to disease should be kept confidential and protected from injudicious disclosure. At the same time, information regarding a person's genetic susceptibilities or diagnoses also provides information about, and may have important implications for, blood relatives who share some of the same genes, a fact that distinguishes genetic information from much other medical information. An ethical dilemma resides in deciding whether patient confidentiality should be protected if providing confidential genetic information to a relative might protect that relative from harm or give him or her reason to alter life choices substantially. Equally troubling is the issue of whether the relative would choose not to have this information (Wertz & Fletcher, 1989; Scanlon & Fibison, 1995).

Using primarily a principle-based framework incorporating elements of an ethic of care, this chapter will explore selected and perplexing ethical issues that are related to genetic testing and services and that affect the individual. Nurses in a variety of care settings will increasingly be called upon to assist their patients in making decisions and to provide care, education, and support throughout the delivery of genetic services.

ETHICAL CONSIDERATIONS IN GENETIC TESTING

As more genes are identified and sequenced through the work of the HGP, development of diagnostic tests for a variety of disorders will increase. Genetic tests are already available for a number of single-gene disorders and are likely to become more accessible as awareness, demand, and profitability expand. Multiplex testing (i.e., multiple genetic tests performed on a single blood or tissue sample) will most likely be available in the near future. Genetic diagnostic testing differs from most other diagnostic tests in that it does not depend on a particular tissue type or clinical state and can be obtained at any stage of life (from conception on). Almost all cells contain DNA, and only a small sample, usually obtained from the blood, is necessary for genetic testing. Although genetic testing can be used to diagnose a disorder in a person who has clinical symptoms, a feature that distinguishes this type of testing from other *diagnostic* testing is that the former also permits diagnosis of the probability that a disorder will occur in an individual who has no symptoms. In these cases of *predictive* testing, the test can predict the likelihood of disease in this individual at some future time, but the timing of onset or the severity of symptoms is largely unpredictable. In addition, genetic tests can identify *carriers* of recessive genetic disorders who themselves will never develop the genetic disease but who could transmit it to some of their offspring. Genetic defects, disorders, and susceptibilities can be detected through testing not only in a given person but also in that person's unborn children, in a fetus, or in a preimplantation embryo.

Much of the current excitement about genetic testing centers around predictive gene testing. Already we can test in research laboratories for nearly two dozen disorders (e.g., Huntington's disease, Alzheimer's disease, and breast cancer). For most genetic disorders, a disturbing gap exists between the ability to diagnose or predict a genetic disorder and the ability to prevent, treat, or alter the course of the disorder. Genetic information—whether diagnostic, predictive, or indicative of carrier status—provides personal information about blood relatives and raises questions about who among them may want to know this information. The intensely personal nature of the information generated by a genetic test, its power to alter major life decisions and to affect family members, and its potential misuse all raise important ethical considerations. Although the technical aspects of genetic testing are becoming less complicated over time, the decisions people must make may be more difficult (Murphy & Lappe, 1994; Scanlon & Fibison, 1995).

Autonomous Decision Making

A fairly wide consensus exists that the decision to undergo genetic testing should be a patient's autonomous decision, made on the basis of that individual's assessment of the risks and benefits and individual values and most often within the context of the clinician-patient relationship. This consensus acknowledges that people have a right to make personal decisions about their own bodies and property, which is consistent with the emphasis on individual liberty in our culture and legal system. A person attempting to decide about testing should always be provided adequate counseling, and fully informed consent should be obtained from the person. Providing adequate information in the appropriate cultural and educational context and allowing for free, unpressured decisions about testing is a demonstration of respect for the autonomy of the individual. Because issues surrounding testing are complex and the consequences are profound, careful discussion and reflection must accompany all decision making. Maintenance of the integrity and dignity of the individual who is seeking testing is paramount, and is best accomplished through a careful and individualized risk/benefit evaluation conducted with the patient. To conduct such an evaluation, the evaluator will need to ensure that the patient has adequate knowledge of the test and its implications and will need to be familiar with the values and interests of the individual. A decision about testing that is autonomous must also be voluntary. A person should be tested only if he or she desires the information and not in response to pressure or undue influence by family members, health care providers, researchers, or others. Discussion between health care professionals and the patient of privacy concerns, the implications for family members, the impact of test results on future choices, possible discrimination, and psychological reactions should all be part of the informed-consent process and pretest counseling. The process and limits of maintaining confidentiality should be discussed (Elias & Annas, 1994; Andrews et al., 1994; Bove et al., 1997).

Testing of Children

Testing of children, especially predictively, remains controversial. Children are not able to give consent for any medical intervention and so, in most cases, their parents or guardians consent for them. To date, the most widespread use of genetic testing is in newborns. Each year in the United States, blood samples are taken from more than 4 million newborn children for testing for missing or abnormal gene products. Testing for phenylketonuria (PKU) is done routinely at birth. Interestingly, the physician or health care team does not routinely seek the consent of the parents to test for PKU but rather honors an explicit refusal of consent when it is expressed. The Institute of Medicine Committee on Assessing Genetic Risks recommends that children not be tested unless the results can lead to direct medical benefits for the child. Parents should be advised to avoid making difficult choices about any other kind of genetic testing for their children but to wait until the children are old enough to choose for themselves (Andrews et al., 1994).

Benefits and Risks

Careful assessment of the anticipated risks and benefits of each proposed test for each patient should precede decisions about the use of diagnostic tests in general, including genetic tests. In genetic testing, the benefit/risk assessment will vary depending on the mutation or defect for which the test is being conducted, the quality and reliability of the assay, the state of knowledge regarding the meaning of any mutations identified, and available courses of action if tests indicate the presence of a genetic mutation or defect (Jeungst, 1996).

In all cases of genetic testing, it could be argued that knowledge is better than uncertainty, even if the knowledge is of a genetic predisposition to disease. Clearly, however, this value of knowledge over uncertainty is not universally held: many at-risk individuals choose not to be tested for disorders, such as Huntington's disease, for which testing is now available (Marteau & Richards, 1996). For some people, uncertainty about their own genetic makeup allows them to maintain hope. Information obtained from genetic testing has implications for the future health and welfare of not only the tested individual but also any children or future children and blood relatives.

As aforementioned, knowledge reduces uncertainty and anxiety for many people and allows for more informed decision making about important life choices. There are other benefits of genetic testing: In some disorders, a genetic test may provide a diagnosis and an explanation of symptoms. In other disorders, such as PKU or hereditary hemochromatosis, testing permits early intervention, which can benefit disease outcomes. For some people, learning that a genetic test is negative not only provides immense relief in most cases but also imparts knowledge that can sometimes be used to reduce the need for costly, invasive, and uncomfortable health-surveillance procedures (e.g., colonoscopy for familial adenomatous polyposis) or interventions (e.g., prophylactic mastectomy for people with a family history of breast cancer and a family pattern of BRCA1 gene). In some disorders, knowledge of a genetic mutation may enable important lifestyle changes that can potentially affect the disease course (e.g., adoption of a high-fiber diet for people with a genetic predisposition to colon cancer). In addition, knowledge provided by genetic tests can assist people in making important life choices. For example, one woman who tested positively for the Huntington's disease gene chose to adopt a child rather than to chance transmitting the gene to her offspring (Marteau & Richards, 1996). People identified as carriers through genetic testing might use this information when making decisions about marriage or reproduction. For example, in some Hasidic Jewish communities, arranged marriages are common practice, and religious objections to abortion and contraception are widely held. In these communities, anonymous but coded testing for the Tay-Sachs gene is performed on women at age 18 and on men at age 20, and the results are entered into a registry. When a marriage is planned, the registry is queried and if both prospective partners are carriers, they are advised to find other partners, thus preventing expression of Tay-Sachs disease (Andrews et al., 1994).

For all its benefits, genetic testing also entails multiple risks. Physical risks from the actual testing are minimal and limited to the discomfort associated with drawing blood. The risks of psychological reactions to the knowledge that one does have a genetic disorder—especially one that is not yet manifest in disease, one for which there is no treatment or prevention, or one that may have already been transmitted to offspring—may include depression, anxiety, despair, a sense of lack of self-worth, anger, and guilt. Strong evidence exists that even news about a negative genetic test may result in severe psychological reactions, such as guilt or depression, especially if other members of the family are affected (Marteau & Richards, 1996). Because information is revealed about others through testing, both the decision to be tested and the results can have a tangible effect on the tested person's family. The fact that some family members may not want to know their risks or to have others know them has an impact on the tested person's decisions and reactions.

A positive genetic test has social risks, also. Stigmatization by family members, friends, or others is possible. The risk of discrimination in health insurance, in jobs, and in other areas of life is real and pervasive. Restrictions on life choices, either self-imposed or prescribed by society, are another risk. Self-imposed restrictions may result in missed opportunities for childbearing, marriage, or career choices, which might later be regretted. Restrictions imposed by others may result in frustration, anger, isolation, and despair. Finally, there is the risk that a patient will make medical choices that have major and long-term consequences, some of which may be potentially dangerous or of unproven value.

Because genetic testing is a rapidly evolving science, some as-yet-unknown risks and benefits of testing likely will become apparent only with further research or after substantial use of a given test. As the Institute of Medicine Committee on Assessing Genetic Risks suggests, until risks and benefits have been defined, "genetic testing and screening programs remain a form of human investigation" (Andrews et al., 1994, p. 4), thus necessitating prior strict scrutiny of anticipated risks and benefits and adherence to the ethical norms and regulations governing research on human subjects.

Privacy and Confidentiality

Genetic information, like other medical information, is considered confidential and should not be disclosed without the tested individual's consent. Genetic information is intensely personal information that people generally wish to keep private, yet protecting the privacy of genetic information is increasingly difficult. DNA testing can be done on stored samples or on blood and tissue samples collected for other reasons. DNA databanks are being established, and genetic disease registries already exist. A patient's consent should always be obtained before DNA testing is performed on identifiable samples or before his or her name is provided to a disease registry or DNA sent to a databank.

Medical information is stored in electronic databases, to which many people have access. The tested individual may have difficulty keeping genetic information private, for example, if questioned by an insurance company or a prospective employer. Such disclosures do not necessarily include results of DNA testing, but family history and the use of genetic services may provide indirect information about genotype that insurance companies may wish to explore further. Additionally, a tested person may feel compelled to share information about genetic tests with family members, health care providers, or others. Genetic testing may reveal undesired or even unsought information, such as misattributed paternity. The meaning and implications of genetic information may be misinterpreted by patients, families, employers, or clinicians. The Institute of Medicine Committee on Assessing Genetic Risks supports a National Society of Genetic Counselors (NSGC) statement on confidentiality, which appears in their 1991 publication *Guiding Principles* (Andrews et al., 1994, p. 157):

> The NSGC support individual confidentiality regarding results of genetic testing. It is the right and responsibility of the individual to determine who shall have access to medical information, particularly results of testing for genetic conditions.

> Patients should be encouraged and supported, however, in sharing appropriate genetic information with spouses and relatives. Through sharing this information, individuals not only demonstrate respect and caring for others but may be giving others information that will allow them to avoid harm or improve outcomes. Because genetic information also has the potential to be misused to deny people employment, insurance, or other social goods or opportunities, we, as a society, should continue to seek ways to decrease the possibility of such discrimination.

Prenatal Testing

Some of the most difficult and personal issues in genetic testing involve prenatal diagnosis. As with most genetic testing, the ability to diagnose conditions prenatally exceeds any ability to treat, prevent, or alter the course of most genetic disorders, yet one of the major applications of genetic information is in reproductive planning and decision making. Several widely used prenatal procedures are designed to provide information about the health and genetic makeup of the fetus; among these are screening of maternal serum alpha-fetoprotein (MSAFP), amniocentesis, chorionic villi sampling, and ultrasonography. Newer, less invasive diagnostic technologies and tests for an increasing number of genetic conditions, including multifactorial disorders, are on the horizon. Despite the increasing ease with which information about genetic conditions can be acquired prenatally, a pregnant woman is generally left with two alternatives if the fetus tests positively for a genetic disorder: either terminate the pregnancy or carry to term an affected child. Selective termination of a pregnancy is beset with the charged debate over the morality of abortion. For people who oppose abortion in principle as destruction of life, selective abortion for a serious genetic condition may not be an option. For people who endorse abortion as a viable option, selective abortion to avoid possible suffering from genetic disease or disability may be a reasonable and welcome option. Before prenatal genetic testing is performed, the pregnant woman or expectant couple should be provided with detailed information about the type, quality, and meaning of available tests and should be given the opportunity to consider and discuss the meaning of such information, their values with regard to terminating a pregnancy or raising a child with a disability, anticipated family reactions and values, cultural and religious influences, and previous experiences. After adequate information and counseling, a pregnant woman should be allowed to choose whether she will undergo prenatal testing. As with all genetic testing, the quality of the test is of utmost importance, because test results provide the basis for serious and irreversible decisions.

Eugenics

An important and controversial societal consideration is whether and to what extent reducing the number of genes responsible for various genetic disorders (sometimes referred to as *negative eugenics*) should be a goal of prenatal diagnosis, any genetic testing that influences reproductive decision making, or potential germline interventions (Munson, 1992). Germline therapy (sometimes referred to as *positive eugenics*) is a means of correcting or eliminating deleterious genes by treating the sperm or ova of a developing embryo as a second, and equally significant, consideration. (See the section on gene therapy.) Some people are concerned that buying into a myth of genetic perfection (i.e., a belief that the perfect family or society, with disease-free genomes, is achievable) will generate intolerance of diversity at the societal level and lead to policies or laws that require or promote selective reproduction. Some also envision an inevitable slide down the "slippery slope" from reducing genetic diseases to actually enhancing or selecting for desirable genetic traits. Also, knowledge of genotypes might allow for logical planning for reducing costs associated with disease and disability in societies that desire to do so (Caplan, 1993).

Both positive and negative selection already occur, as people elect not only to terminate pregnancies to *eliminate* a genetic disorder such as Down syndrome but also, in some cases, to *select* for a specific gender. The consensus of the Instutite of Medicine Committee on Assessing Genetic Risks (and others) is that prenatal diagnosis should be offered only for the diagnosis of genetic disorders and birth defects. Society may already exert great influence on certain selection decisions—for example, by refusing health insurance to a person or families with a given genetic disorder. People have been denied insurance, employment, and loans on the basis of genotype alone (Billings et al., 1992). Continued public debate about these issues is critical and inevitable.

GENETIC SERVICES AND GENE THERAPY

Because treatment for genetic disorders is often unavailable, currently most genetic services consist of diagnosis and counseling. Referral to ongoing research study groups is also possible. Gene therapy, although in its infancy, offers hope for the future.

Counseling

People should be appropriately informed about their genetic risks and any diagnostic, behavioral, or therapeutic options available for dealing with those risks. Individuals who otherwise might be unaware of their options can use the information provided to choose among the alternatives of conventional treatments and strategies and newly developed genetic technologies and to make critical life decisions. *Genetic counseling* "refers to the communication process by which individuals and their family members are given information about the nature, recurrence risk, burden, risks and benefits of tests, and meaning of test results, including reproductive options of a genetic condition, as well as counseling and support concerning the implications of such genetic information" (Andrews et al., 1994, p. 4). Although genetic counseling has been primarily conducted by a relatively small and well-trained group of geneticists and genetic counselors, such counseling will increasingly be the responsibility of primary care physicians, nurses, social workers, and other front-line health care providers. Unlike genetic counselors, these professionals are likely to have had little formal training in genetics and to practice a generally directive style in their interactions with patients. Education of health care providers and the lay public about genetic discoveries and their clinical application is essential. Sensitivity to the modes and bases for different styles of counseling will be an important part of health care provider education.

Counseling should always be conducted with respect for the autonomy and privacy of the individual(s), informed consent, and protection of confidentiality. Genetic counselors have generally adopted a philosophy and style of nondirectiveness in counseling. The goal is to facilitate decision making, to help clients to reach a carefully considered decision but not to tell them what to decide. This effort entails support for the individual's right to make voluntary informed decisions about testing, follow-up, reproductive options, and the like, in addition to individualized and compassionate support for the emotional turmoil caused by such decisions and their outcomes. Although genetic counseling is an interactive and engaging process, the counselor refrains from making a recommendation and leaves the decision in the hands of the counselee.

Nondirectiveness is based on values such as freedom and individual autonomy. Counseling cannot be totally value free, however (Caplan, 1993). Those people involved in genetic services inevitably bring to bear on their professional pursuits their personal values and perspectives, which can influence their presentation or description of a disorder or its outcomes. It has been shown that decisions can be influenced not only by the presentation of information but also by who presents it (Marteau & Richards, 1996).

Many professionals believe that nondirectiveness should remain the standard for reproductive planning and decision making, because freedom of choice over reproduction has a high value in Western culture. Freedom to choose is not always welcome, however, and can be accompanied by high psychological costs (e.g., anxiety and guilt). Informed refusal of genetic counseling must also be respected.

The ethics of genetic counseling warrants continual reexamination, especially in light of the expanding number, type, and availability of genetic tests. More controversial is whether nondirectiveness is the appropriate counseling style when predictive tests for disease susceptibility are being offered and patients are being assisted through the complexities of a positive test result. For genetic disorders with treatment possibilities, counseling should resemble more the provision of other medical advice regarding treatment. In cases in which strategies (preventive, therapeutic, or behavioral) exist that are believed to be beneficial to the individual, these options should be recommended and encouraged. Further research is needed to evaluate the benefit of interventions that might delay onset or diminish the severity of a predicted genetic disease. Further research and dialogue are needed regarding different modes and outcomes of genetic counseling for different purposes and in various settings.

Gene Therapy

Gene therapy is a method of treating genetic disorders by manipulating the genes themselves. The two main classifications of gene therapies are somatic and germline. *Somatic gene therapy* involves the correction of genes within somatic cells in specific identifiable organs or tissues of individual patients. The goal is to treat or cure a genetic condition that is modifiable by the insertion of a specific gene or the elimination of a gene function. During the last 5 or so years, many investigational protocols of gene therapies for single-gene diseases (such as cystic fibrosis and adenosine deaminase deficiency) and for cancer or the human immunodeficiency virus were initiated (Ross et al., 1996). Nonetheless, the vast majority of these are first-phase studies having a goal of establishing safety. Efficacy has not yet been proven for any somatic gene therapy, and no such therapy has been approved as standard treatment for disease (King et al., 1992).

Because the goal of somatic gene therapy is treatment of a particular patient, the major related ethical consideration is similar to that of any treatment offered to an individual patient; that is, the expected benefits to the patient should outweigh any risks or discomforts to the patient. Careful attention to ethical issues and protecting human subjects (in terms of safety, informed consent, and confidentiality) in clinical trials of somatic gene therapies is also necessary. The Recombinant DNA Advisory Committee (RAC) of the National Institutes of Health (NIH) serves an important role in public deliberations and consideration of ethical concerns in gene therapy. In experimental uses of somatic gene therapy, the guiding principles, as in all human-subjects research, demand a favorable benefit/risk evaluation, the informed consent of the subject, protection of a subject's rights and welfare, and fairness in subject selection.

Germline gene therapy, on the other hand, involves treating a preimplantation embryo using in vitro fertilization techniques or treating the germline cells (ova or sperm) of adults so that genetic defects are not transmitted to their children. Because germline therapy is intended to treat future generations as well as individual patients, it is morally different from other types of therapy. Germline therapy affects cells in which all or most of an individual's genetic information is yet to be activated and used, thereby introducing a new element of uncertainty and possible risks to the future child or his or her children. Development of such techniques requires research on embryos, which currently is not permitted using U.S. federal funds. Embryo research might involve selecting out and discarding unhealthy embryos or fetuses, which is considered by some people to be morally untenable. Concern also centers on the possibility that in correcting serious genetic disorders in today's population through germline therapy, scientists may cause even more serious maladies in future generations. Germline therapy to correct or eliminate deleterious genes is also seen as the first step down the slippery slope into improvement or enhancement of human traits, sometimes referred to as *positive eugenics.* The ethics of germline gene therapy are very complex and will require continued public debate and reflection. At this time, there is a general moratorium on the use of such therapy.

Challenges and Implications

The clinical application of genetic knowledge brings together ethical issues normally confronted in caring for people in ordinary clinical practice and ethical issues particular to health care practices oriented to groups and populations (e.g., human-subjects research or public health). Perhaps unique to genetics, the "group" may be the patient's family, potential beneficiaries of medical research, the public at large, or even future generations. The consequence can be ethical synergy when the interests of the individual and group coincide, and ethical tension when they conflict. In the care of individuals, difficult ethical decisions and dilemmas regarding genetic testing or the provision of genetic services require reflection, dialogue, compassion, and principled and reasoned decision making. Likewise, difficult decisions and dilemmas associated with the application of new genetic knowledge that affects families, societies, and future generations require careful thought, public dialogue, systematic study, reasoned decision making, and a commitment to determining decisions and policies of which we will all be proud.

THE ETHICAL, LEGAL, AND SOCIAL IMPLICATIONS PROGRAM OF THE HGP

The HGP has set aside approximately 5 percent of its budget for the study of the ethical, legal, and social implications (ELSI) of the work of the Project. The goal of the ELSI Program of the National Center for Human Genome Research (NCHGR) is to identify and develop initial responses to the most urgent ethical, legal, and social issues posed by genome research. Unique in the history of science, the ELSI program "gives us the opportunity to 'worry in advance' about the implications and impacts of the mapping and sequencing of the human genome, including several thousand human disease genes, *before* wide-scale genetic diagnosis, testing, and screening come into practice rather than *after* the problems have presented themselves in full relief" (Andrews et al., 1994, p. 3). The ELSI program supports research projects, conferences, working groups, fellowships, and other initiatives that have focused on such issues as privacy and confidentiality, health provider preparedness and education, discrimination in insurance and employment, quality control of DNA testing procedures, and public education. The ELSI program and its early products provide preliminary evidence that supports the idea that research on and public deliberation of the uses of new knowledge will aid in the well-informed evolution of social policies about science.

CONCLUSION

Genetic discoveries are presenting possibilities for diagnosis and treatment of health conditions and health options that did not exist previously. These possibilities bring with them ethical, social, and legal considerations in individual, institutional, and societal realms. Nurses are becoming increasingly involved in offering genetic information and testing to patients and families and in recording and reporting genetic information. They will therefore need to develop the ability to carry out careful assessment of the anticipated risks and benefits of genetic information and testing for each patient and his or her family and to ensure that genetic information is handled with respect for privacy and confidentiality.

The four principles that are central to biomedical ethics—respect for autonomy, nonmaleficence, beneficence, and justice—provide a framework within which nurses can consider ethical issues related to genetic information and testing. Privacy, confidentiality, discrimination, access to and justice in health care, and informed decision making, although not new, are particularly pertinent issues in the field of genetics. These issues, when considered within the context of an ethic of care that is founded on an understanding of each person's uniqueness and the role of relationships, emotions, and character traits in moral deliberations and judgments, provide ethical guidance for all nurses who care for patients and families with genetic concerns. Having an understanding of the relevance of ethics to genetics will help guide nurses in searching for and appraising solutions to ethical problems.

SUMMARY POINTS

Relevance of ethics

- Availability of new possibilities and choices as a result of scientific advances and with them uncertainty about the ethical, or proper, thing to do
- Ethical dimensions of choice, a component in decision making about application of scientific knowledge.
- Need to consider ethical decisions in different realms—individual, institutional, and societal

Approaches to ethics

- Ethics: a branch of philosophy concerned with values related to human conduct, the rightness or wrongness of certain actions, and the goodness or badness of the motives or ends of such actions
- Four principles central to biomedical ethics
  - respect for autonomy
  - nonmaleficence
  - beneficence
  - justice
- Additional important ethical principles
  - fidelity
  - confidentiality
  - respect for people

- Ethic of care: takes into consideration the uniqueness of individuals, relationships, and their value, emotions, and character traits, as these play important roles in moral deliberations and judgments

Ethical issues commonly confronted in genetics

- Challenging ethical issues in genetics
  - privacy
  - confidentiality
  - discrimination
  - prenatal testing
  - abortion
  - access to and justice in health care
  - informed consent
  - patient decision making
- Patient confidentiality and implications for blood relatives

Ethical considerations in genetic testing

- Diagnostic genetic testing: not dependent on a particular tissue type or clinical state and can be obtained at any stage of life
- Predictive genetic testing: can be used for diagnosis and permits identification of the probability of a disorder in a person without symptoms
- Carrier testing: can identify carriers of genetic traits
- Autonomous decision making: acknowledges the person's right to make personal decisions about his or her own body and property
  - voluntary decision making, without coercion
  - possible discrimination
  - psychological issues
- Testing of children
  - somewhat controversial
  - current recommendations: no testing of children unless results can lead to direct benefit
- Benefits of genetic testing
  - reduces uncertainty and anxiety through knowledge
  - provides explanation
  - permits intervention
  - reduces cost of care

- Risks of genetic testing
  - increased anxiety, depression, and despair
  - feelings of lack of self-worth, anger, and guilt
  - stigmatization
  - employment and insurance discrimination
  - restriction of life choices
- Privacy and confidentiality
  - personal and private nature of DNA information
  - need for consent for release of information or tissue
  - concerns regarding insurance and employability
  - concerns regarding the release of genetic information to family members and health care providers
  - revelation of unsought information (i.e., paternity)
- Prenatal testing
  - difficult personal decisions in genetic testing for prenatal diagnosis include acquiring information prenatally, and having to decide whether to terminate a pregnancy or to continue
  - women/couples should be offered, prior to prenatal testing, detailed information about the type, quality and meaning of available testing; have the opportunity to consider and discuss the meaning of such information in light of family values, cultural and religious influences, and previous experiences
  - importance of test quality
- Eugenics
  - negative eugenics: use of prenatal and other genetic testing to reduce the number of genes responsible for various genetic disorders
  - "myth of genetic perfection": concern that genetic testing will be used to create the perfect family or society and generate intolerance of diversity at a societal level
  - logical plan for reducing costs associated with disease and disability; possible in light of knowledge of genetic genotypes
  - need for continued public discussion and debate to address the potential for insurance and employment discrimination

Genetic services and gene therapy

- Counseling
  - "a communication process by which individuals and their family members are given information about the nature, recurrence

risk, burden, risks and benefits of tests, and meaning of test results, including reproductive options of a genetic condition, as well as counseling and support concerning the implications of such genetic information" (Andrews et al., 1994, p. 4)

- benefit of nondirective approach for reproductive planning and decision making: allows for freedom of choice over reproduction, which is valued in Western cultures
- possible drawbacks of nondirective approach: may be inappropriate counseling style when offering predictive tests and assisting people through the complexities of positive test results, assuming beneficial treatments are available

- Gene therapy
  - a method of treating genetic disorders by manipulating the genes themselves
  - goal of gene therapy: to treat or cure a genetic condition that is modifiable by insertion of a specific gene or elimination of gene function
  - two main classifications: somatic and germline
  - ethical issues associated with somatic gene therapy: benefit/risk ratio, safety, informed consent, confidentiality, protection of human subjects in clinical trials
  - ethical issues associated with germline gene therapy: current proscription of research and treatment of embryos, possibility that correction of serious genetic disorders might lead to other, more serious health concerns for future generations, need for further public debate

- Ethical challenges and implications of application of genetic knowledge
  - practice oriented to groups and populations such as human-subjects research or public health
  - some groups unique to genetics: the patient's family, potential beneficiaries of medical research, the public at large, and future generations
  - need for nurses to employ careful reflection, dialogue, compassion, and reasoned decision making when caring for patients, families, and the broader community

The Ethical, Legal, and Social Implications program of the Human Genome Project

- 5 percent of HGP budget
- goal is to identify and develop initial responses to the most urgent ethical, legal, and social issues
- supports research projects, conferences, and working groups to address such issues as privacy and confidentiality, discrimination, and quality control of DNA-testing procedures

QUESTIONS FOR CRITICAL THINKING

1. Genetic discoveries are viewed by some as positive and by others as opening a Pandora's box. Consider some examples of each perspective. How would you respond to patients who voice concerns? Discuss both sides of this issue.

2. How would you propose that nurses achieve competence in facilitating personal and professional bioethical decision making when integrating genetics into practice?

3. Choose for discussion a challenging genetics case example from your clinical setting. Explore with nursing colleagues and other health professionals potential solutions to any ethical dilemmas found within this situation. Use the questions in Table 9.2 to guide your problem solving.

REFERENCES

Andrews, L., Fullarton, J., Holtzman, M., and Motulsky, A. (eds.) (1994). *Assessing genetic risks: implications for health and social policy.* Washington, DC: National Academy Press.

Beauchamp, T., and Childress, J. (1994). *Principles of biomedical ethics* (4th ed.). New York: Oxford University Press.

Bergem, A. L. M. (1994). Hereditary dementia of the alzheimer type. *Clin. Genet.* 46:144–149.

Biesecker, B. (1997). Psychological issues in cancer genetics. *Semin. Oncol. Nurs.* 13(2):129–134.

Billings, P., Kohn, M., de Cuevas, M., et al. (1992). Discrimination as a consequence of genetic testing. *Am. J. Hum. Genet.* 50:476–482.

Bove, C. M., Fry, C. T., and MacDonald, D. J. (1997). Presymptomatic and predisposition genetic testing: ethical and social considerations. *Semin. Oncol. Nurs.* 13(2):135–140.

Caplan, A. (1993). Neutrality is not morality: the ethics of genetic counseling. In Bartels, D. M., LeRoy, B., and Caplan. A. L., eds., *Prescribing our future: ethical challenges in genetic counseling,* pp. 149–165. New York: Aldine de Gruyter.

Delieu, J., and Keady, J. (1996). The biology of Alzheimer's disease. *Br. J. Nurs.* 5(3):162–168.

Devettere, R. (1995). *Practical decision making in health care ethics.* Washington, DC: Georgetown University Press.

Elias, S., and Annas, G. (1994). Generic consent for genetic screening. *N. Engl. J. Med:* 330(22): 1611–1613.

Gert, B., Berger, E., Cahill, G., et al. (1996). *Morality and the new genetics: a guide for students and health care providers.* Sudbury, MA: Jones and Bartlett.

Jeungst, E. (1996). Respecting human subjects in genome research: a preliminary policy agenda. In H. Vanderpool (ed.). *The ethics of research involving human subjects,* pp. 401–429. Frederick, MD: University Publishing Group.

Jonsen, A., and Toulmin, S. (1988). *The abuse of casuistry.* Berkeley: University of California Press.

King, R. A., Rotter, J. I., and Motulsky, A. G. (1992). *The genetic basis of common diseases.* New York: Oxford University Press.

Lynch, J. (1997). The genetics and natural history of hereditary colon cancer. *Semin. Oncol. Nurs.* 13(2):91–98.

Marteau, T., and Richards, M. (1996). *The troubled helix: social and psychological implications of the new human genetics.* Cambridge, Engl.: Cambridge University Press.

Munson, R. (1992). *Intervention and reflection: basic issues in medical ethics* (4th ed.). Belmont, CA: Wadsworth.

Murphy, T., and Lappe, M. (eds.) (1994). *Justice and the Human Genome Project.* Berkeley: University of California Press.

Nelson, D. L. (1993). Fragile X syndrome: review and current status. *Growth Genet. Horm.* (2):1–15.

Noddings, N. (1984). *Caring: a feminist approach to ethics and moral education.* Berkeley: University of California Press.

Reamer, F. (1991). *AIDS and ethics.* New York: Columbia University Press, 1991.

Ross, G., Erickson, R., Knorr, D., et al. (1996) Gene therapy in the United States: a five-year status report. *Hum. Gene Ther.* 7:1781–1790.

Scanlon, C., and Fibison, W. (1995). *Managing genetic information: implications for nursing practice.* Washington, DC: American Nurses Association.

Shapira J. (1994). Research trends in Alzheimer's disease. *J. Gerontol. Nurs.* 20(4):4–9.

Wertz, D., and Fletcher, J. (eds.) (1989). *Ethics and human genetics: a cross-cultural perspective.* New York: Springer-Verlag.

CASE STUDY 9.1

AUTONOMY IN DECISION MAKING: FAMILIAL DEMENTIA

Mary Sue, age 28, visits her obstetrician, because she has concerns related to her family history. While taking a detailed family history, Mary Sue's nurse learns that several maternal aunts and uncles and her maternal grandfather suffered from dementing conditions. It is also believed that comparable disorders were present in previous generations on the patient's mother's side, but elder family members do not want to discuss the details of these conditions.

Mary Sue has recently read newspaper articles suggesting that Alzheimer's disease* has a genetic component and that in one family the relevant gene was discovered. Mary Sue tells the nurse that she and her siblings, ages 24–40, are at odds. Several siblings, including Mary Sue, would like a genetic workup to determine whether they are at risk for developing Alzheimer's disease. Their mother is opposed to any type of testing. She and the remainder of the family are afraid of the

Alzheimer's disease is the most common cause of dementia affecting patients over 65. Patients who have Alzheimer's disease experience progressive impairment in memory, language, orientation, judgment, and personality. These changes are progressive and exceed those generally associated with age. Alzheimer's disease is generally classified as being of either early or late onset. Early-onset disease occurs in patients who develop symptoms before age 60, whereas late-onset disease occurs in patients who develop symptoms after age 60. Diagnosis is based on clinical symptoms as defined by the National Institute of Neurology Cognitive Disease Standards. Currently, a definitive diagnosis can be made after a brain biopsy or an autopsy provides clinical and histopathological evidence.

A family history of Alzheimer's disease might increase a person's risk for developing the condition, particularly if the patient is a first-degree relative or if multiple family members are affected. Genetic testing may be helpful for some families, but there are limitations to the testing, such as informativeness of test results and the fact that there is, as yet, no specific treatment or cure. Such testing is therefore currently carried out in special genetics and research settings (Bergem, 1994; Delieu & Keady, 1996; Shapira, 1994).

possible ramifications of knowing that they may have inherited susceptibility that could lead to Alzheimer's disease. They fear, also, that job security and access to affordable health insurance might be jeopardized.

- As the nurse caring for Mary Sue, what is your duty as regards her request for genetic testing for Alzheimer's disease?
- How might you respond to the family's expressed concerns?
- What are some of the ethical issues involved with testing other family members?

Discussion

One can easily understand the mixed emotions of family members described in this case. The older family members may be resistant to discussions of possible familial Alzheimer's disease, because they feel guilt about the prospects of possibly having passed along such an illness to their offspring. They may feel anxious about developing the condition themselves, as their life trajectory moves them closer to average age of onset. Further, their resistance may be influenced by social patterns set earlier on, making difficult any open discussion of family illnesses.

One can understand, also, the discord among the siblings. Given that Alzheimer's is a disease of relatively late onset, for which no specific treatments and no prospect for cure exist, some siblings see no reason to know whether this disease will affect them. Others might want to know in order to reduce the anxiety produced by uncertainty. Still others might cope by avoiding grappling with the information altogether. Resistance associated with concerns about employability and insurability are grounded in societal concerns raised by genetic testing.

Such family disagreements are commonplace and can be addressed by the health care professional in a variety of ways, depending on the professional's own values and ethical preferences. If the nurse interprets his or her responsibilities as extending to the patient only, the nurse would feel he or she had a duty to the identified patient or research subject alone: that is, only when the nurse has formally entered into relationship with a patient do such professional duties attach.

If a nurse interprets his or her duties and obligations in terms of the patient only, he or she will consider it acceptable to assist only Mary Sue in exploring the possibility of the testing she seeks. Such ethical interpretation removes from the nurse the need to consider concretely the potential benefits and harms that such information could bring to other family members. Though such a professional might be willing to interact with Mary Sue's family members, these interactions would result only from arrangements initiated and organized by Mary Sue.

If, on the other hand, a nurse believes that testing would be not only in Mary

Sue's best interest but also clinically beneficial for other family members and believes that he or she has responsibilities to both the patient and significant others (i.e., sees the family in a more unitary sense), he or she might take a more active role in bringing some consensus to the family's discussions. For example, in collaboration with the health care team, the nurse might invite the family members to come to a meeting facilitated by the nurse.

Determining whether the patient, Mary Sue, is best served by interpreting nursing duties and obligations to the patient alone or to the extended family may be influenced by the nurse's assessment of Mary Sue's reasons for seeking this information and what the information might provide her. Given our lack of scientific understanding of the meaningfulness of genetic testing in Alzheimer's disease, coupled with the lack of specific interventions and cure, the prudent course for now might be to discuss with Mary Sue these concerns regarding genetic testing for this disease.

CASE STUDY 9.2

PRIVACY AND CONFIDENTIALITY: MUST I TELL MY SISTER ABOUT MY SON'S FRAGILE X CONDITION?

Four-year-old Ryan is brought to the medical genetics clinic by his 27-year-old mother, Janet, for evaluation of developmental delay and hyperactivity. Janet tells the nurse that she hopes that they can find the cause of Ryan's delays, because she and her husband, Terry, are hoping to have other children, and they want to know whether their other children will have similar problems.

The nurse notes that the family history is unremarkable, with the exception of a 6-year-old son of one of Janet's cousins, who is apparently "slow." Janet reports that her family members do not get along, and she has had little contact with her mother's side of the family. Her parents are both deceased. A friend told Janet recently that Janet's 25-year-old sister, Linda, has just learned that she is pregnant with her first child.

It is determined that the cause of Ryan's delay is fragile X syndrome* and that Janet is a carrier. After discussing the genetics of fragile X syndrome and its implications for

Fragile X syndrome is the most common form of inherited mental retardation, occurring in both males and females. Patients who have fragile X syndrome may, in addition to varying degrees of mental retardation, exhibit behavioral problems such as hyperactivity and typical physical features, including a long, narrow face and protruding ears. The range of severity of clinical symptoms varies;

other family members, Janet is asked to notify her sister. The following week, the nurse calls Janet, who reports that she hasn't yet called her sister.

- As the nurse, what is your obligation to Janet?
- How might you discuss with Janet the issue of notifying Linda and other family members?
- Do you have an obligation to inform Linda?

Discussion

One of the nurse's obligations to Janet is to ensure that she has an opportunity to discuss and review the implications of the diagnosis of fragile X syndrome for her son, her reproductive future, and her family. If, as in the preceding case, the nurse views his or her relationship as being limited to the patient—in this case, Janet—the nurse might feel no obligation to the patient's family, especially an estranged family. That

some patients (both male and female) carry the gene for fragile X syndrome but display no symptoms. Fragile X syndrome is caused by a gene mutation on the X chromosome. Diagnostic and prenatal testing are available and may be offered to families (Nelson, 1993). (See Chapter 6 for a more detailed discussion of fragile X syndrome.)

the two conditions are different—that is, Alzheimer's disease is a late-onset disorder and fragile X syndrome is congenital—does not alter or increase the provider's responsibilities. In both conditions, there is neither a specific treatment nor a cure.

Clinical symptoms of fragile X syndrome range from none to moderate impairment, and prediction of the outcome in utero is not always possible. In the nurse's role as educator and advocate for Janet and her husband, the nurse will want to ensure that the couple is given appropriate information and counseling, so that Janet and her husband can make a fully informed decision about future reproduction and prenatal testing.

As a provider who is unconnected to Linda, the nurse can have no idea of Linda's emotional state. Sharing this genetic information may do more harm than good. Thus, although he or she has an obligation to discuss with Janet the value of informing Linda, the professional is under no moral obligation to ascertain that the genetic information is conveyed to Janet's sister, once the topic has been raised. The nurse could simply remind Janet that he or she and the health care team are always available to answer further questions and to assist Janet and her family.

C ASE S TUDY **9.3**

ASSESSMENT OF ANTICIPATED
BENEFITS AND RISKS OF GENETIC
TESTING: TESTING MINORS FOR
CANCER SUSCEPTIBILITY

Sarah is a 40-year-old woman with colorectal cancer* that has metastasized to
her liver. She is currently receiving experimental therapy and shares with the nurse
and physician that her father had colon
cancer, her sister has polyps, and she is
concerned about possible genetic explanations. She is referred to the Cancer
Genetics Clinic.

On assessment of her family history, the
nurse learns that Sarah has a sister, Susan,
34, from whom three suspicious polyps
have been removed. Her father was diagnosed with colon cancer when he was 48
years old and was treated successfully with
surgery but now has recurrent colon cancer
at age 65. Sarah's greatest concern at this
time is for her two sons, ages 10 and 15.

*It is estimated that 10 percent of patients who
have colorectal cancer (CRC) are members of families with an inherited autosomal-dominant form
of CRC. Genes responsible for certain types of
hereditary CRC, such as familial adenomatous
polyposis, have been identified, making it possible
to test for CRC to confirm a suspected diagnosis
and for predisposition and prenatal diagnosis.
Genetic testing for CRC may benefit families with
hereditary CRC by allowing for early detection
and prevention and improved surveillance and
management strategies (Biesecker, 1997; Bove et
al., 1997; Lynch, 1997).*

Her elder son has had two polyps
removed, and she is extremely anxious
over his potential for developing cancer.
Sarah very much wants to have herself and
her sons tested to determine whether they
carry any of the now-known genes for
colon cancer.

Sarah reports that her sister is now ter-
rified that she is going to get colon cancer
and has decided that she doesn't want to
know anything else about the disease. She
especially does not want to know whether
Sarah or her sons carry any genes predisposing them to colon cancer.

- What are some of the ethical concerns
  involved in testing Sarah's two sons?
- How might you, as the nurse, explain
  and discuss these concerns with Sarah?
- How would you discuss Sarah's sister's
  response? What is your obligation to
  the sister?

Discussion

In his or her discussion with Sarah and her
physician, the nurse could consider with
Sarah the current issues regarding genetic
testing of minors. In general, the current
practice is not to test minors for adult-
onset diseases, such as colon cancer.
Although this position may be overridden
in conditions that have reported pediatric
onsets, for which there are effective intervention or surveillance strategies and that
have the potential for cure, one changes
this position cautiously. Although the con-

ditions necessary for considering a change in this position might be met in the case of familial colon cancer, the natural history of the specific colon cancer that affects the family could influence approaches to diagnosis and interventions without using genetic testing.

Rarely, documented cases of colon cancer have been diagnosed in people younger than 20 years in high-risk families. A diagnosis of colon cancer in people younger than 30 is also uncommon, but some argue that if a person is diagnosed in her or his early twenties, that person may have already had the disease for a few years. Possibly, the disease could have been detected earlier, but knowledge of one's genetic-mutation status is not necessary to accomplish this. Early colonoscopy may have detected such a cancer.

Current guidelines suggest initiating colonoscopy at age 25 in high-risk families or at 5 years younger than the earliest age at diagnosis in the kindred. Given the medical facts in Sarah's family to date, it would be wise to start closely monitoring her sons. The necessary monitoring can be initiated on the basis of family history without benefit of genetic tests.

Given that good medical management of Sarah's sons calls for close monitoring, one should carefully consider the benefits and risks posed by performance of the genetic test. Having this test result on file might mean insurance or employment discrimination. Knowing one's status may affect self-image, education, and reproductive decisions. If the balance of benefits and burdens is uncertain, providers should respect the decision of competent adolescents and their families. If the medical or psychosocial benefits of a genetic test will not accrue until adulthood, as in the case of carrier status or adult-onset diseases, genetic testing generally should be deferred.

The health care team does not have a specific obligation to Sarah's sister, Susan, at this point. The team can encourage Sarah to share information with her sister and can offer their availability to both if they desire further information. Even if Susan does not want to be tested, there are surveillance issues for her (in terms of frequency of colonoscopy, etc.) about which she could be educated. If Susan is in a patient-physician relationship, her physician could encourage her to increase her surveillance based on her family history alone, regardless of genetic-mutation status. Susan can implement the surveillance measures of a gene-mutation carrier without ever having to know her own status.

It is the health care team's responsibility to convey an accurate picture of the potential benefits and burdens of whatever decisions are made. For example, potential physical consequences include increased physical surveillance, which can be costly and uncomfortable, or decreased surveillance if one's test result is negative for a known gene mutation in his or her family.

Potential emotional effects that Sarah might experience include anxiety or distress related to confirmation of mutation status, although a greater measure of certainty can reduce or alleviate anxiety for some. Family relationships may be affected positively or negatively. Members of a family who test negatively may experience survivor guilt, whereas those who test positively may experience "transmitter guilt" if they believe or know that they have passed a gene mutation to offspring.

Potential social effects include a plethora of discrimination: insurance problems, employability, social stigmatization. Some mutation carriers may opt to pay for extra screening out of pocket so as not to alert their insurance carrier. This can be very costly. Conversely, knowledge of genetic mutation status has created staunch advocates of some patients who are at risk. Carriers of genetic mutations are actively advocating for insurance reform and an agenda that addresses the needs of this population for increased education of their providers and of the general public. Predicting which of these possible results might occur for Sarah or Susan is difficult, if not impossible. Thus it is incumbent on the nurse and the health care team to ensure that all involved are adequately counseled before any final decisions are reached.

Future Directions

One day Alice came to a fork in the road and saw the Cheshire cat in a tree. "Which road do I take?" she asked. His response was a question: "Where do you want to go?" "I don't know," Alice answered. "Then," said the cat, "it doesn't matter."

Lewis Carroll

OBJECTIVE
Describe future roles for nurses in genetics.

RATIONALE
Many unanswered questions remain that are related to genetic information and its successful integration into clinical practice, research, and public policy. Knowledge of these unresolved issues will prepare nurses to participate in developing educational methods, clinical applications, and social policies to meet the needs of patients, families, and the broader community.

APPLICATION ACTIVITIES
• Plan for continuing education in genetics.
• Become familiar with genetics and nursing informatics resources.
• Participate in educational activities that will increase bioethical decision-making skills.

GENE CARE LINK

For more information about issues to consider for the future, go to http://www. jbpub.com/ clinical-genetics.

INTRODUCTION

T he science and technology evolving from the efforts of the Human Genome Project (HGP) will provide many opportunities for new diagnostic and treatment applications in health care. The four fruits that this Project bears are knowledge, tools, opportunity, and obligation. Applications gained from this information-gathering expedition will require an understanding of the meaning of such information to the individual in terms of diagnosis, intervention, health promotion, and prevention of illness. The translation of scientific knowledge into clinical applications presents many challenges (Rogers, 1995). The information gleaned from the HGP will be employed to dissect the intricacies of the human body, particularly in the development of health and disease. Exploration of the repair of damage caused by hereditary, environmental, dietary, and other factors will naturally flow from the applications of the HGP discoveries. All health care providers will have to assume an active role in using these genetic advances, which will present both opportunities and obligations. This chapter continues a process intended to help the nurse consider issues and challenges for nursing practice in genetics. From these fundamentals, the nurse then must consider a world in which genetics knowledge is applied effectively in the clinical setting, and should contemplate the accommodations that must be made in the future to ensure that that world becomes a reality. Information presented builds on concepts and content previously presented, and offers a foundation from which the nurse can project his or her upcoming responsibilities to patients, families, and the broader community.

KNOWLEDGE

What do the discoveries of the HGP offer as a foundation for a future concept of health care? The first fruit of all labor is knowledge gained: genetic-linkage maps, physical maps, and gene-sequence information. As greater understanding is gained of human genes and their locations on the human chromosomes, more specific knowledge of normal and abnormal genetic function will result. Yet, even as this knowledge is gained, additional questions will arise (Neel, 1997). The incidence of and correlation between genetic changes, disease development, and health promotion, for example, remains to be determined. Understanding what happens in the human body when a disease gene or mutation is present will provide opportunities for further exploration of the cause, treatment, and cure of genetic conditions.

Many questions at the level of genes, proteins, and cellular functions have yet to be answered. How do genetic changes occur? What happens as a result of a genetic change? Are there ways in which the body corrects these changes? Why does this correction of gene damage work for some but not all genes? How does a change in one gene affect other functions within the body? As health care professionals, we are all challenged to promote and participate in the design of clinical research to help address these questions.

In the future, diagnosis by direct detection of genetic mutations will become increasingly available. Considerations for future genetic testing include the situations in which to offer testing for carrier detection, prenatal diagnosis, and presymptomatic diagnosis of such complex conditions as cancer. (Issues related to our understanding of cancer genetics and the current state of cancer-risk assessment and counseling were discussed in Chapter 8.) How useful is the diagnostic test for detecting or predicting a genetic condition? How sensitive and specific is the genetic test? Can it acquire the necessary information and can we be assured that the results obtained are valid and relate to clinical outcomes? What populations and communities are interested in using genetic testing? How will access to genetic testing be assured and through what payment methods? Should access to genetic testing ever be restricted (e.g., in the case of children)? If so, who should formulate such restrictions? These issues, as well as the cultural, social, and religious backgrounds of patients, families, and communities might influence the application of existing and developing technology.

Another focus for future genetic technology will be to assess the most appropriate treatments for patients, based on the underlying genetic causes of disease. Individualized prevention or treatment strategies may be selected on the basis of basic biological phenomena. Our understanding of the clinical course and outcome of genetic conditions, both in those who carry gene mutations and in those who do not, in response to specific interventions, may provide guidance for health promotion, interventions, and treatments for the general population. Can all patients, for example, who have the same genetic condition be treated in the same manner or are patient-specific plans of care based on an individual's genetic profile more effective?

TOOLS

The tools that facilitate our understanding of the genetic code and the basis of health and disease are continuously advancing and improving. The cost, in time and resources, of methods and techniques for mapping and sequencing the human genome, for example, is decreasing. Access to equipment, skilled personnel, and funding for such basic and clinical research as is being undertaken by the HGP is essential for successful application and translation of genetic technologies in the commercial market.

The computer, for example, has provided a means of enhancing genetic research. Continuous communication of discoveries through a common database has enabled researchers to make use of others' discoveries and has thereby facilitated ongoing research. This sharing among investigators, while it may challenge territorial boundaries, also stretches those boundaries through collaboration. Unresolved issues related to computer tools include ways to develop an interactive yet secure data-informatics system that facilitates the exchange of information but protects the confidentiality of data.

Additionally, computer literacy is a skill that all researchers will need to develop and update to maximize the use of the computer in this arena.

Other tools being developed include educational and counseling methods and materials for patients and families who have genetic concerns and conditions. Interactive tools to enhance both patient and health-care-provider education are needed. Assessment tools, analysis methods, criteria for identification of those at risk for genetic conditions, and psychosocial measurement tools are some of the materials needed to facilitate the integration of genetics into clinical practice. A clearinghouse of genetics resource materials is one mechanism that would enhance sharing of such educational tools.

OPPORTUNITY

An important consequence of the HGP is that all health care providers now need to know about basic genetic concepts and their application to practice. The opportunity for each discipline to delineate the scope of practice and the role of genetics in health care settings is exciting. Yet, to provide knowledgeable and competent care in genetics, nurses must receive a broader education in genetics. As more is learned about the genetic component of common conditions such as cancer (see Chapter 7), nurses will have opportunities to integrate genetic concepts into their current practice skills and knowledge bases. Another opportunity for nurses in the future is the establishment by schools of nursing of educational curricula that incorporate genetics (Curry & Wergin, 1993).

Research efforts focused on educating nurses can help to identify educational needs and to field test and implement the most effective educational models. Areas of research in nursing education might include identification of nursing roles in the provision of genetic health care, nursing preparedness for specific genetic-counseling tasks, efficacy of models of training and education in preparing nurses for expanded roles in genetic testing, development of methods for adapting aspects of genetic counseling to nursing care, and assessment of the impact of incorporating genetic principles and practice into nursing care.

Developing models of care that integrate genetic and other specialists and primary health care providers offers another opportunity for enhancing collaboration and continuity of care and for preventing duplication of effort in the clinical setting. Genetic discoveries affect all disciplines, across many specialties. Translational methods for basic and applied research that encompass all disciplines must be explored. What kind of counseling clinics, for example, might best meet the needs of patients, families, and communities? Who is qualified to provide genetic testing, counseling, and education? What are the most effective ways to empower patients to make informed health care decisions?

Professional nursing societies are playing an active role in preparing members to meet the demands of translating genetic discoveries into nursing practice. With the American Medical Association and the American Nurses Association, organizations representing many health care professionals and consumers are exploring the opportunity to join forces in a national coalition for health professional education in genetics. Professional nursing societies that adopt such a proactive role in preparing for the future are exhibiting visionary leadership and a willingness to invest in the future.

The opportunity for nurses to assume leadership and to educate others (e.g., policy makers and the public) will require ongoing incorporation and translation of genetic discoveries into nursing practice. Public educational methods, the Internet, videotape, CD-ROMs, and written materials are all media through which nurses can develop and share their expertise. As the public becomes aware of some of the implications of genetic research, an opportunity and a responsibility are created for thoughtful and appropriate education.

OBLIGATIONS

As noted by McKusick in 1997, two scientific and technological revolutions are converging. The biological revolution and the information revolution are providing access to tremendous amounts of information about humans. This information carries with it many obligations for all health care professionals, to ensure the appropriate use of the knowledge, tools, and opportunities made available by the HGP and by future discoveries.

One obligation is leadership. A national agenda should be created that addresses ethical, legal, and social issues resulting from our ability to understand and integrate genetic technology into health care. Health care professionals, especially nurses who are accustomed to caring for families and communities, will participate in genetic research and applications of genetic knowledge to practice. They are therefore responsible for providing leadership to the nation in developing policies that reflect the concerns of the public (Kanungo & Mendonca, 1996). Policy development will continue to be a challenge in the future as new information becomes available. Balancing the rights to privacy, free choice, and ownership of genetic information, for example, are considerations that must be addressed. The government's role in ensuring nondiscrimination, quality control, and health care coverage are other examples of policy issues that must be determined.

At this juncture, the preparation of all people—patients, families, communities, and health care providers—for taking part in bioethical decision making is critical. Eventually, all of us will be faced with complex decisions; to make such decisions, we will need to possess sufficient knowledge to balance alternatives for health care. Although patient autonomy in health care decisions has previously been encouraged, the potential now exists for greater patient involvement in health care decisions, particularly regarding how, when, and whether genetic information will be used in their health care and possibly in that of future generations. Nurses must begin to assess their personal values and beliefs about the contribution of genetics to health and illness and to contemplate their experiences with genetic conditions. Only then will awareness of the potential impact of such influences as personal circumstances, ethical beliefs, and socioeconomic, religious, and psychosocial backgrounds become integrated into nurses' understanding of patient decision making. Future educational considerations must include the development and evaluation of effective methods for teaching bioethical decision making, to ensure the appropriate use of genetic information by individuals and by society (Jones & Beck, 1995; Parker, 1994).

CONCLUSION

Genetic discoveries are being made at an ever-increasing rate, with nearly daily announcements of new genes and new technologies and their implications. The increasing ability of scientists and medical specialists to understand and manipulate the human genome has the potential to generate public fears and concerns. These concerns are not entirely unfounded, as social history provides examples of the misuse and abuse of genetic information. Nurses have a responsibility to participate in the successful integration of genetic information and methods into all health care settings. Especially important is the role of nurses in patient advocacy. Nurses are also expected to provide educated leadership in the application of genetics to clinical practice. These roles will create exciting personal and professional challenges.

SUMMARY POINTS

Impact on Nursing Practice: The impact of expanding genetic knowledge on nursing practice is limitless. Many opportunities and challenges exist as health care professionals strive to integrate into current clinical practice all that is known about the genetic contribution to health and disease, in order to improve and enhance health care.

Nursing Issues: Clinical trials are necessary to determine the usefulness of genetic information for improving outcomes of care. The nursing-care models that will most effectively provide patient access to genetic information and health care options, the types of resources necessary for providing this care, and the best ways to prepare patients for genetic information have not been identified. Many unanswered questions remain. For example, does possession of genetic information affect prevention, detection, and management decisions? Nurses are challenged to mobilize their personal and professional resources to provide leadership in answering such questions.

Self-Awareness: Take time out to reflect on, refresh, and reframe how you visualize the future. For instance, how do you envision helping others to integrate genetic information into their daily lives? How do you plan to prepare for the challenges of including genetic knowledge in clinical practice? How might you help other professionals to prepare? In what ways can you participate in efforts to ensure that genetic discoveries are used to improve the health and well-being of society?

People attempting to answer such questions will need to hone their bioethical decision-making skills. Many of the ethical issues of genetics research and practice are global. Rapid scientific advances and the technological achievements of the Information Age are contributing to worldwide discussion of such issues. Nonetheless, emotions, personal beliefs and values, and cultural diversity may engender discomfort and conflict around these issues. Nurses can help to balance the positive and negative aspects of this evolving area of patient care through advocacy, education, and research, and thereby enhance outcomes for the health of all.

QUESTIONS FOR CRITICAL THINKING

1. Describe the ways that you envision genetic discoveries influencing the future direction of patient care.

2. Discuss your concerns about and hopes for the future when contemplating genetic influences on society's choices.

3. Contemplate your potential personal and professional roles in integrating genetics into your life and the lives of others.

REFERENCES

Curry, L., Wergin, J., and Associates (1993). *Educating professionals: responding to new expectations for competence and accountability.* San Francisco: Jossey-Bass.

Jones, R., and Beck, S. (1995). *Decision making in nursing.* Albany, NY: Delmar.

Kanungo, R., and Mendonca, M. (1996). *Ethical dimensions of leadership.* Fayetteville, AR: Sage.

McKusick, V. (1997). History of medical genetics. In D. Rimoin, J. Connor, and R. Pyeritz (eds.), *Emery & Rimoin's principles and practice of medical genetics* (3rd ed.). Edinburgh: Churchill Livingstone.

Neel, J. (1997). Coming events casting shadows ahead. In D. Rimoin, J. Connor, and R. Pyreitz (eds.), *Emery-Rimoin principles and practice of medical genetics* (3rd ed.). Edinburgh: Churchill Livingstone.

Parker, L. (1994). Bioethics for human geneticists: models for reasoning and methods for teaching. *Am. J. Hum. Genet.* 54:137–147.

Rogers, E. (1995). *Diffusion of innovations* (4th ed.). New York: Free Press.

Quality Time

The blue sky above
The white clouds below
Insecure, anxious, and unsure of which way to go
Thinking of opportunities, challenges, work to be done
But fear and concern for the future
and I'm only one.

But one can make a difference
A team can do even more
So let's join together and begin to explore.
Today's the beginning
A chance to brainstorm
A chance to work together
A chance to do more.

You're part of this transition
You'll design a plan
To change the world vision
Of health and illness for man.
So let's design the future
One thing certain, nothing's for sure!
We can test the boundaries,
and let nurses soar.

Making a difference
Showing the way
Integrating genetics into living
Tomorrow's a new day.

Jenkins, 11/96

Each of the following case studies offers you the opportunity to consider and apply to your practice genetic principles and information. Questions at the end of each case are intended to stimulate critical thinking and further discussion about your roles in caring for patients and families with current and future genetic health concerns.

CASE STUDY A.1

NEURAL-TUBE DEFECTS

You are a nurse in a prenatal clinic. Ms. N, an 18-year-old college freshman, comes in for her first prenatal visit at 16 weeks of pregnancy. While you are taking the family and medical history, Ms. N explains that this is an unplanned pregnancy. She also tells you that her boyfriend's brother died within the past year, at age 19, from complications associated with spina bifida. She says that her boyfriend was very close to his brother and that this has been a difficult year for him. You discuss with Ms. N the alpha-fetoprotein (AFP) profile screening test for neural-tube defects and Down syndrome. Ms. N says that she wants to have the test but states that she doesn't know what she will do if her baby has a neural-tube defect. She explains, "My boyfriend would probably want to keep the baby, but I am afraid of how I would handle all of the responsibilities."

A week later, Ms. N's AFP profile screening test results come back showing an increased risk for a neural-tube defect in her fetus. Her doctor calls and explains the results to her and makes a referral to a high-risk clinic for ultrasound examination and genetic counseling. You review with Ms. N the purpose and goals of genetic counseling and explain that the ultrasound examination can give more detailed information about the presence of a neural-tube defect in her fetus.

You learn from the genetic counselor later that day that Ms. N and her boyfriend brought their parents with them to the counseling session and that they talked at length about Ms. N's boyfriend's brother and spina bifida. The counselor explained that during the session, Ms. N did not say very much.

The ultrasound examination revealed a large, open, abdominal-wall defect and some heart abnormalities but no neural-tube defect. The counselor tells you that amniocentesis was offered to Ms. N and her boyfriend and that they decided to undergo this test to learn more about whether a chromosomal abnormality might be present. You try to contact Ms. N at home to follow up on her visit, but her mother tells you that she is not taking any telephone calls.

Results of the amniocentesis reveal trisomy 18, a chromosomal condition that involves an extra chromosome number 18. The doctor offers Ms. N follow-up genetic counseling to discuss the chromosomes and her options at this point. You learn that trisomy 18 is a serious condition that causes physical and mental disabilities and that babies born with this condition do not generally survive beyond the newborn period.

Ms. N returns to the clinic the following day to talk with her doctor about the results. When you see her, she is alone. She says, "My boyfriend didn't want to come with me today. He is having a hard time handling this." She also informs you that she had thought about spina bifida and what she would do, but now she is very confused about how to proceed, as she did not expect this news. She tells you that she is considering all of her choices at this point. She says that she is afraid to make a decision that will alienate her boyfriend and her parents because of their religious beliefs.

Questions

1. What are some of the genetic counseling issues for Ms. N? for her boyfriend? for the families?

2. How could you provide the most support to Ms. N in this situation?

3. What additional resources might you identify to support Ms. N, her boyfriend, and their families?

CASE STUDY A.2

CYSTIC FIBROSIS

Mr. P is a 21-year-old patient in a health maintenance organization. During his annual company physical examination, he informs you that he has a sister, age 19, who has cystic fibrosis. He also had a sister who died shortly after birth from complications associated with cystic fibrosis. He tells you that he is getting married within the next year and that he and his fiancé want to obtain information about the likelihood of his being a carrier of a gene for cystic fibrosis and their having children with cystic fibrosis.

Mr. P explains that his sister had DNA testing for cystic fibrosis, and he knows that this information will be useful if he decides to undergo carrier testing. He tells you, "The only problem is that my sister and I don't get along, so I don't know whether she would be willing to share this information with me." Mr. P tells you that he and his mother are very close and that he has talked with her about his concerns about being a cystic fibrosis carrier. Mr. P asks you to help him get the information

he needs to learn more about his carrier status.

Questions

1. Explain how the results of Mr. P's sister's tests would be helpful in determining Mr. P's carrier status.

2. What resources could you offer to Mr. P to help him to get answers to his questions? How would you describe these resources?

3. Describe some of the potential ethical and counseling issues for Mr. P and his family.

MYOTONIC DYSTROPHY

Mr. and Mrs. F have a 2-year-old son who has developmental delays. The primary care physician in your clinic refers the couple for genetic evaluation and counseling. You explain to Mr. and Mrs. F what they can expect, including a physical examination of their son and a written summary of the evaluation. Mrs. F tells you that she hopes that they will find a solution to her son's problems, because she is 40 years old and wants to have more children. She says she wants to know the likelihood of having other children with the same problems.

You receive and review the summary of the F's genetic-counseling session. You learn that a condition called *myotonic dystrophy* is suspected of being the cause of Mr. and Mrs. F's son's delays and that DNA testing is being done to confirm this diagnosis. You also learn that the genetics specialists wonder whether Mrs. F also has a mild form of myotonic dystrophy. Several weeks later, you receive the results of the DNA testing, which confirm the diagnosis of myotonic dystrophy. Mr. and Mrs. F return to the genetics clinic to obtain further information about the diagnosis.

The nurse specialist at the genetics clinic calls you in follow-up to Mr. and

Mrs. F's visit. She explains to you that Mr. and Mrs. F are glad to have information about the cause of their son's delays, because it will help them to plan appropriately for his care and education. However, Mrs. F was upset to learn that she may also have myotonic dystrophy. She is reported to have told the genetics nurse specialist, "I wasn't prepared for you to tell me this! I thought we were here to talk about my son and my future pregnancies!"

The genetics nurse specialist tells you that Mrs. F is not convinced that she should have DNA testing for herself. She is worried about her insurance. During the counseling session, Mrs. F stated, "I don't know what to do about this information. I just wasn't prepared, and now I don't know what to do about a future pregnancy."

Questions

1. What are some of the genetic counseling issues for Mr. and Mrs. F?

2. How might you respond to Mrs. F's concerns about her own diagnosis and about future family planning?

3. How would you respond if Mrs. F asked, "What would you do if you were me?"

4. What are some of the possible reproductive issues for Mrs. F if she learns that she does have myotonic dystrophy?

Title	*Source*
American Society of Human Genetics (ASHG) Policy Statement for Maternal Serum Alpha-Fetoprotein Screening Programs and Quality Control for Laboratories Performing Maternal Serum and Amniotic Fluid Alpha-Fetoprotein Assays	*Am. J. Hum. Genet.* 1987, 40:75–82
American Society of Human Genetics Social Issues Committee Report on Genetics and Adoption: Points to Consider	*Am. J. Hum. Genet.* 1991, 48:1009–1010
American Society of Human Genetics on Clinical Genetics and Freedom of Choice	*Am. J. Hum. Genet.* 1992, 48:1011
American Society of Human Genetics Position Paper on Patenting of Expressed Sequence Tags	ASHG, Nov. 1991; available on-line at http://www. faseb.org/genetics/ashg/ policy/pol-00.htm
ASHG Human Genome Committee Report: The Human Genome Project: Implications for Human Genetics	*Am. J. Hum. Genet.* 1991, 49:687–691
DNA Banking and DNA Analysis: Points to Consider	*Am. J. Hum. Genet.* 1988, 42:781–783
Genetic Testing and Insurance	*Am. J. Hum. Genet.* 1995, 56:327–331
National Action Plan on Breast Cancer Position Paper on Hereditary Susceptibility Testing for Breast Cancer	*J. Clin. Oncol.* 1996, 14:1738–1740
Proposed ASHG Position on Mapping/Sequencing the Human Genome	*Am. J. Hum. Genet.* 1988, 43:101–102
Presymptomatic Genetic Testing for Heritable Breast Cancer Risk	National Breast Cancer Coalition, 1995; available on-line at http://wwwicic. nci.hih.gov/_genetics.html

Title (cont'd.)	*Source (cont'd.)*
Recommendations for Follow-Up Care of Individuals with Inherited Predisposition to Cancer (HNPCC)	*J.A.M.A.* 1997, 277:915–919
Recommendations for Follow-Up Care of Individuals with Inherited Predisposition to Cancer (BRCA1 and BRCA2)	*J.A.M.A.* 1997, 277:997–1003
Statement of the American Society of Clinical Oncology (ASCO) Genetic Testing for Cancer Susceptibility	*J. Clin. Oncol.* 1996, 14:1730–1736
Commentary on the ASCO Statement on Genetic Testing for Cancer Susceptibility	*J. Clin. Oncol.* 1996, 14:1737–1740
Statement of the American Society of Human Genetics on Cystic Fibrosis Carrier Screening	*Am. J. Hum. Genet.* 1992, 51:1443–1444
Statement of the American Society of Human Genetics on Genetic Testing for Breast and Ovarian Cancer Predisposition	*Am. J. Hum. Genet.* 1994, 55:i–iv
Statement of the American Society of Human Genetics on Points to Consider: Ethical, Legal, and Psychosocial Implications of Genetic Testing in Children and Adolescents	*Am. J. Hum. Genet.* 1995, 57:1233–1241
Statement on Informed Consent for Genetic Research	*Am. J. Hum. Genet.* 1996, 59:471–474
Statement on the Use of Apolipoprotein E Testing for Alzheimer's Disease	American College of Medical Genetics; available on-line at http://www.faseb.org/genetics/ashg/policy/pol-00.htm
Statement on Use of DNA Testing for Presymptomatic Identification of Cancer Risk	*J.A.M.A.* 1994, 271:785
Testing Recommendations for Germline p53 Mutations in Cancer-Prone Individuals	*J. Natl. Cancer Inst.* 1992, 84:1156–1160
Update of MSAFP Policy Statement from the American Society of Human Genetics	*Am. J. Hum. Genet.* 1989, 45:332–334

Patient Consumer Information

Alliance of Genetic Support Groups
35 Wisconsin Circle
Suite 440
Chevy Chase, MD 20815-7015
1-800-336-GENE
301-652-5553
e-mail: alliance@capaccess.org

(a coalition of voluntary organizations and professionals that provides
resources for a variety of genetic conditions for patients, families, and
professionals)

National Resources

National Institute for Human Genome Research (NHGRI)
NIH Bldg. 31, Room 4B09
Bethesda, MD 20892
301-496-0844
http://www.nhgri.nih.gov

Ethical, Legal, Social Implications Branch (NHGRI)
NIH Bldg. 38A, Room 617
Bethesda, MD 20892
301-402-4997

National Cancer Institute
NIH Bldg. 31, Room 10A29
Bethesda, MD 20892
301-496-6631
1-800-345-3300
http://www.nci.nih.gov

National Institute for Nursing Research
NIH Bldg. 31, Room 5B03
Bethesda, MD 20892
301-496-8230
http://www.ninr.nih.gov

Regional Genetics Networks

Council of Regional Networks for Genetic Services (CORN)
Medical Genetics/Pediatrics
Emory University
2040 Ridgewood Drive
Atlanta, GA 30322

Genetics Network of the Empire State, Puerto Rico
   and the Virgin Islands (GENES)
Wadsworth Center for Laboratories and Research
Room E-299
P.O. Box 509
Albany, NY 12201-0509
518-486-2215
518-474-8590 (fax)

Great Lakes Regional Genetics Group (GLaRGG)
University of Wisconsin
328 Waisman Center
1500 Highland Avenue
Madison, WI 53705-2280
608-265-2907
608-263-3496 (fax)

Great Plains Genetics Service Network (GPGSN)
Division of Medical Genetics
Department of Pediatrics
The University of Iowa
Iowa City, IA 52242-1038
319-356-4860
319-356-3347 (fax)

Mid-Atlantic Regional Human Genetics Network (MARHGN)
Family Planning Council
260 South Broad Street, Suite 1900
Philadelphia, PA 19102-3865
215-985-2600

Mountain States Regional Genetic Services Network (MSRGSN)
Colorado Department of Health
FCHS-MAS-A4
4300 Cherry Creek Drive South
Denver, CO 80222-1530
303-692-2423
303-782-5576 (fax)

New England Regional Genetics Group (NERGG)
P.O. Box 670
Mt. Desert, ME 04660-0670
207-288-2704
207-288-2705

Pacific Northwest Regional Genetics Group (PacNoRGG)
Oregon Health Sciences University
Child Development and Rehabilitation Center
P.O. Box 574
Portland, OR 97207-0574
503-494-8342
503-494-4447

Pacific Southwest Regional Genetics Network (PSRGN)
Genetic Disease Branch
2151 Berkeley Way, Annex 4
Berkeley, CA 94704-9802
510-540-2696
510-540-2095 (fax)

Southeastern Regional Genetics Group (SERGG)
Emory University School of Medicine
Pediatrics/Medical Genetics
2040 Ridgewood Drive
Atlanta, GA 30322-4870
404-727-5844
404-727-5783

Texas Genetic Network (TEXGENE)
Bureau of Maternal and Child Health
Texas Department of Health
1100 West 49th Street
Austin, TX 78756-3199
512-458-7700
512-458-7421 (fax)

Professional Genetics Societies

International Society of Nurses in Genetics (ISONG)
Eileen Rawnsley, Executive Director
7 Haskins Road
Hanover, NH 03755
603-643-3028

National Society of Genetic Counselors (NSGC)
233 Canterbury Drive
Wallingford, PA 19086-6617
610-872-7608
e-mail: BEANSGC@aol.com

American Society of Human Genetics (ASHG)
9650 Rockville Pike
Bethesda, MD 20814-3998
301-571-1825
301-530-7079 (fax)

American College of Medical Genetics (ACMG)
9650 Rockville Pike
Bethesda, MD 20814-3998
301-530-7127
301-571-1895 (fax)

NATIONAL CENTER FOR HUMAN GENOME RESEARCH

F · A · C · T · S · H · E · E · T

G E N E T I C
M A P P I N G

A p r i l   1 9 9 6

FINDING A SINGLE GENE IN the huge morass of DNA that makes up the human genome--some 3 billion base pairs of it--requires a set of powerful tools. The Human Genome Project is developing three basic types of tools to make gene hunts and DNA analysis faster, cheaper, and practical for almost any scientist to do. These tools include genetic maps (also called linkage maps), physical maps, and DNA sequence--a detailed description of the order of the nucleotide bases in DNA. Developing new technologies to provide better and better tools is an ongoing goal of the Human Genome Project.

Genetic mapping is the first step in isolating a gene. It offers firm evidence that a disease or trait is linked to the transmission of one or more genes from parent to child. At the same time, genetic mapping provides clues about which chromosome contains the gene and precisely where on that chromosome the gene lies. Genetic maps have been used successfully for several years to seek out single genes responsible for inherited disorders. With steady improvements on the maps, they have become more useful in guiding scientists to the many genes that interact to bring about more complex, every-day disorders, such as asthma, heart disease, diabetes, cancer, and psychiatric illness.

Genetic mapping begins with collecting blood or tissue samples from members of families in which a disease or trait is prevalent. DNA is isolated from these samples and examined for characteristic molecular patterns, or markers, that are inherited by family members along with an inherited disease.

Even before researchers have identified the gene actually responsible for a trait, markers tell them roughly how close by the gene is. That's because of a genetic process called "recombination." During the development of eggs and sperm, each pair of chromosomes in those cells mix and exchange (recombine) genetic material. If a particular gene is close to a marker, the gene and marker will likely stay together during recombination and be passed on together to a child. If they are far apart, the odds are low that the gene and marker will be passed on together. So, if each family member who has a particular trait or a disease also inherits a particular marker, chances are high that the gene responsible for the disease lies close to that marker. Consequently, knowing where the marker is also tells a scientist roughly where the gene is. The more markers there are on the map, the more likely one will be closely linked to a disease gene, and the easier it will be to zero in on that gene.

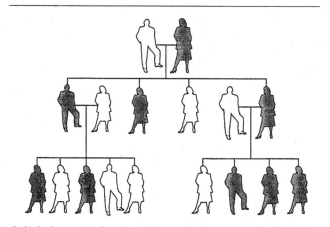

In this family tree, or "pedigree," a tendency to develop breast cancer is inherited. Children of a parent who carries a mutated BRCA1 gene (shown in gray) have a 50:50 chance of also inheriting the mutated gene. Pedigree information, plus the development of reliable DNA markers, helped researchers track the gene and determine that it is located on chromosome 17.

An early major goal of the Human Genome Project was to develop dense maps of markers spaced evenly throughout the genome. With increasing quality of the markers and higher density on a chromosome, genetic mapping can now help a researcher localize almost any gene. In 1994, the human genetic map became the first of the major goals of the Human Genome Project to be reached. At that time, a large, international group of investigators published a comprehensive genetic map of the human genome containing nearly 6,000 markers. With markers spaced less than 1 million bases apart on average, this map is more detailed than the goal originally called for. What's more, this goal was reached a full year ahead of schedule. Subsequently, scientists have continued to put finishing touches on the genetic map, but that phase of the Human Genome Project is now essentially complete.

Markers consist of slight differences in the arrangement of letters in the genetic alphabet--A, T, C, and G--on individual chromosomes. Because these differences, called polymorphisms, usually occur in DNA that does not contain a gene, they don't usually affect a person's health. But they can tell a researcher which person the DNA came from, which makes them extremely valuable in tracking inheritance of traits through several generations of a family. They are also useful for forensic applications.

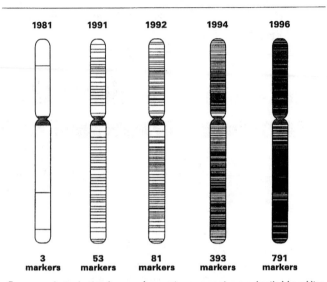

Genome project scientists have made genetic maps much more detailed by adding hundreds of new markers to each chromosome. The markers on this genetic map of chromosome 4 help scientists pinpoint the location of a gene.

Although there are several different types of genetic markers, the type most used on genetic maps today is known as a microsatellite. These markers are of high quality and easy to use with automated laboratory equipment, so mapping a trait in a large number of family members can be done rapidly.

The development of high-resolution, easy-to-use genetic maps promises to revolutionize genetics research. Using current genetic maps, an investigator can rapidly localize a suspect gene to a chromosomal region "just" a few million base pairs in length. The improved quality of genetic linkage maps has reduced the time required for such gene mapping from a period of years to, in many cases, a matter of months or weeks.

Genetic mapping data generated by Human Genome Project laboratories is stored in databases scientists can freely access. In the United States, the Genome Database (GDB, http://gdbwww.gdb.org/gdb/docs/gdbhome.html) serves as a public repository for mapping information.

NCHGR Office of Communications, 31 Center Drive, Building 31, Room 4B09, MSC 2152, Tel: 301.402.0911, FAX: 301.402.2218
http://www.nchgr.nih.gov

F a c t s   a b o u t   G e n e t i c   M a p p i n g

NATIONAL CENTER FOR HUMAN GENOME RESEARCH

F   A   C   T   S   H   E   E   T

P H Y S I C A L
M A P P I N G

A P R I L   1 9 9 6

FINDING A SINGLE GENE IN the huge morass of DNA that makes up the human genome-- some 3 billion base pairs of it-- requires a set of powerful tools. The Human Genome Project is developing three basic types of tools to make gene hunts faster, cheaper, and practical for almost any scientist to do. These tools include genetic maps (also called linkage maps), physical maps, and better technologies for determining the order, or sequence, of the DNA base pairs, a process commonly called DNA sequencing.

After scientists use genetic mapping to assign a gene to a relatively small area on a chromosome, they must next examine that region up close to learn the gene's precise location. To do this, scientists turn to physical maps, which provide gene hunters with DNA representing known segments of a chromosome. Physical maps also provide the raw material for DNA sequencing.

To construct a physical map, scientists snip apart a chromosome (or in some cases, the whole genome) into DNA pieces they can work with, and copy, or "clone," the pieces in the laboratory. Then they match up the clones to reflect the same order they were on the original chromosome. Doing this tells researchers exactly where each piece came from, so information about the location and genetic content of the piece will correspond to the actual chromosomal region. Then information about the order is stored in a computer, and

copies of the pieces are stored in laboratory freezers. When genetic linkage mapping has indicated a gene lies in a particular region, scientists can retrieve the copied piece corresponding to that region, and begin looking through it for the gene.

Having the chromosome pieces in the correct order on the physical map is essential to a successful gene hunt and a major challenge to genome scientists constructing physical maps. So one goal of the Human Genome Project is to develop a system of markers that accurately connects cloned DNA pieces, and to do this for the entire genome. The result is a set of cloned DNA pieces that overlap in places that share the same marker. As new DNA pieces are

added, the length of the set extends, until it spans a large region of a chromosome, or even a whole chromosome.

A type of marker commonly used for ordering clones to make the physical map is known as the STS, for "sequence-tagged site." The Human Genome Project aims to produce physical maps with an STS marker on average every 100,000 bases of human DNA. Overall, such a map will require about 30,000 STS markers. Genome scientists have already placed some 15,000 STSs on the physical map and will likely reach 30,000 in the next two years.

Gene hunters will use these markers as mile posts, much like those on an interstate highway, to tell them how close they are to

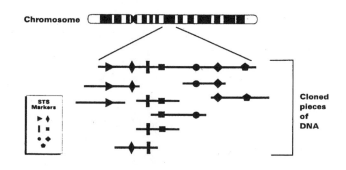

After scientists have cloned large amounts of DNA from the genome, they use markers known as STSs to match up the cloned pieces. The object of physical mapping is to create complete sets of cloned DNA pieces that span to every region of a chromosome and even the whole human genome.

their destination, that is, to the disease gene they are looking for. Not so long ago, with very few markers to guide them, scientists spent years or even a decade traveling along a chromosome in search of a gene. At this level of detail, a skilled researcher should be able to pinpoint the exact location of any gene within a short distance of an STS marker in a fairly short time. And, because some STSs will come from markers on the genetic linkage map, the markers will serve to connect information from the two kinds of maps.

When the project began in 1990, its planners established a goal calling for the development of overlapping clone sets at least 2 million bases long by 1995. Today, these so-called "contigs" are many times longer than that-- ranging from 20 million to 50 million DNA bases in length. In the near future, over 95 percent of the human genome will be covered by overlapping DNA clone sets, each of which is at least 10 million bases long. Already, cloned DNA sets have been completed for chromosomes 21, 22, and Y, and nearly complete sets

have been developed for chromosomes 3, 4, 7, 11, 12, 16, 19, and X. The ultimate goal, of course, is to make cloned DNA sets representing the entire human genome.

Physical maps generated by Human Genome Project laboratories are stored in databases scientists can freely access. In the United States, the best sources of physical mapping data are maintained by the individual laboratories themselves and are available on the Internet:

http://www.nchgr.nih.gov /otherresources/genetic.html

NCHGR Office of Communications, 31 Center Drive, Building 31, Room 4B09, MSC 2152, Tel: 301.402.0911, FAX: 301.402.2218
http://www.nchgr.nih.gov

F a c t s   a b o u t   P h y s i c a l   M a p p i n g

NATIONAL CENTER FOR HUMAN GENOME RESEARCH

F   A   C   T   S   H   E   E   T

P O S I T I O N A L C L O N I N G

A P R I L   1 9 9 6

NOT SO LONG AGO, SCIENTISTS had to have some idea of the biochemical defect in a cell--usually a missing or altered protein--before they could search for the genetic basis of a disease. But in the 1980s, a new technique called positional cloning began to revolutionize the field of human gene mapping.

Taking advantage of recent advances in genetic engineering, positional cloning allowed scientists to map disease-linked genes to a specific chromosome, knowing little beforehand about the disease itself. All they needed to find an altered gene were plenty of DNA samples from families with a history of a disease, powerful analytical tools, and luck.

To understand positional cloning, imagine an astronaut peering down from space trying to locate his son in a park on Chicago's North Shore. Unable to spot the boy with his naked eye, the astronaut picks out landmarks that lead him to the park. He recognizes the shape of North America, then moves in to the outline of Lake Michigan, the Sears Tower, and so on. Once he's zeroed in on the North Shore, the astronaut can rely on more advanced equipment to navigate him to his son.

Like the astronaut, positional cloners begin their search by peering down on all 3 billion base pairs, the chemical building blocks of DNA. Their first step is to find out which of the 24 different human chromosomes contains the altered gene. To do this, they rely on a process called genetic mapping. It involves tracking a known DNA sequence, called a genetic marker. In several family members, genetic variation allows scientists to compare DNA segments, looking for markers shared by those who have the disease. Once they locate a marker that is tightly linked to the disease, like finding Chicago's North Shore, they can focus their gene search on a specific region of a chromosome. From there, the scientists apply sophisticated physical mapping techniques to zero in on and isolate their target.

Since 1986, when scientists found the gene for chronic granulomatous disease--the first disease gene to be isolated by positional cloning--the technique has been used to find well over 50 disease-linked genes. Many more will be found in the future.

The isolation last year of the ataxia telangiectasia (A-T) gene is a prime example of positional cloning success. In brief, here's how the gene was found.

Genetic map - In 1988, after analyzing DNA samples from families with a history of A-T, a research team

---

Disease Genes Identified by Positonal Cloning

1986	1993	1995
Chronic Granulomatous Disease	Menkes Disease	Spinal Muscular Atrophy
Duchenne Muscular Dystrophy	X-Linked Agammaglobulinemia	Chondrodysplasia Punctata
Retinobastoma	Glycerol Kinase Deficiency	Limb-Girdle Muscular Dystrophy
	Adrenoleukodystrophy	Ocular Albinism
1989	Neurofibromatosis Type 2	Ataxia Telangiectasia
Cystic Fibrosis	Huntington Disease	Alzheimer's Disease
	Von Hippel-Lindau Disease	(Chromosome 14)
1990	Spinocerebellar Ataxia 1	Alzheimer's Disease
Wilms Tumor	Lissencephaly	(Chromosome 1)
Neurofibromatosis Type 1	Wilson Disease	Hypophosphatemic Rickets
Testis Determining Factor	Tuberous Sclerosis	Hereditary Multiple Exostoses
Choroideremia		Bloom Syndrome
	1994	Early Onset Breast Cancer (BRCA2)
1991	McLeod Syndrome	
Fragile X Syndrome	Polycystic Kidney Disease	**1996**
Familial Polyposis Coli	Dentatorubral Pallidoluysian Atrophy	Friedreich's Ataxia
Kallmann Syndrome	Fragile X "E"	Progressive Myoclonic Epilepsy
Aniridia	Achondroplasia	Treacher Collins Syndrome
	Wiskott Aldrich Syndrome	Long QT Syndrome (Chromosome 11)
1992	Early Onset Breast/Ovarian Cancer	Barth Syndrome
Myotonic Dystrophy	(BRCA1)	Simpson-Golabi-Behmel Syndrome
Lowe Syndrome	Diastrophic Dysplasia	
Norrie Syndrome	Aarskog-Scott Syndrome	
	Congenital Adrenal Hypoplasia	
	Emery-Dreifuss Muscular Dystrophy	
	Machado-Joseph Disease	

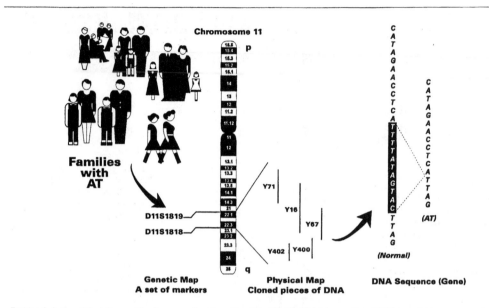

Positional cloning of the A-T gene. The discovery of an eleven base pair deletion (right) in an affected child proved the correct gene had been found.

reported it had found a strong marker that mapped the A-T gene to chromosome 11. The marker was in a region identified as 22 and 23 on the long arm (designated "q") of the chromosome. Further work narrowed down the position of the A-T gene to a region of 3 million base pairs within 11q22-23.

Additional studies led to a higher density map of short sequence linkage markers, like roadsigns, interspersed throughout the 3 million base-pair candidate region. From there, an international consortium of scientists studying 176 A-T families refined the location of the gene as being between two markers, D11S1818 and D11S1819. These markers spanned a distance of about a half-million base pairs.

Physical mapping and isolation of A-T gene - With their search now tightly focused on a very small region of chromosome 11, the scientists reconstructed a physical map consisting of DNA fragments spanning the candidate region. They then mapped cloned gene transcripts to discover the location of genes residing in this region. The scientists were particularly interested in a 20,000 base-pair subregion of Y67, part of the reconstructed DNA fragments, that was statistically most likely to contain the A-T gene. Several genes mapped to this area. One large gene, which stretched nearly 6,000 base pairs, became the primary candidate as the cause of the disease. Additional work led to the identification of an alteration in the gene, now called ATM, which

was strong evidence that the gene causes A-T. To confirm their findings, the researchers went back and tested A-T patients. They found mutation in 14 patients.

As successful as gene searches have become, a new approach called "positional candidate" cloning is rapidly coming online and should streamline the process of positional cloning within the next few years. Positional candidate allows researchers to combine linkage information about a gene's chromosomal location with increasingly detailed gene maps on file in computer databases. These maps, being produced by the Human Genome Project, should greatly accelerate the future discovery of disease-linked genes.

NCHGR Office of Communications, 31 Center Drive, Building 31, Room 4B09, MSC 2152, Tel: 301.402.0911, FAX: 301.402.2218
http://www.nchgr.nih.gov

F a c t s   a b o u t   P o s i t i o n a l   C l o n i n g

NATIONAL CENTER FOR HUMAN GENOME RESEARCH

F A C T S H E E T

D N A
S E Q U E N C I N G

A P R I L 1 9 9 6

FINDING A SINGLE GENE IN the huge morass of DNA that makes up the human genome--some 3 billion base pairs of it--requires a set of powerful tools. The Human Genome Project is developing three types of tools to make gene hunts faster, cheaper, and practical for almost any scientist to do. These tools include genetic maps (also called linkage maps), physical maps, and DNA sequence--a detailed description of the order of the nucleotide bases in DNA. Indeed, a major goal of the Human Genome Project is to sequence the entire length of human DNA. Developing new technologies to provide better and better tools is an ongoing goal of the Human Genome Project.

Knowing the sequence of bases-- the chemical building blocks that make up the DNA strand--is important because it tells scientists what kind of genetic information the DNA carries in that particular segment. Scientists also use sequence information to sort out which stretches of DNA contain genes and to analyze genes for changes in sequence that may cause disease.

The first methods for sequencing DNA were developed in the mid-1970s. At that time, scientists could sequence only a few base pairs per year, obviously not enough to take on a single gene, much less the entire human genome. When the Human Genome Project began in 1990, few laboratories had sequenced even 100,000 bases, and the cost of doing so was more than

$5 per base pair. Since then, technology improvements and automation have increased speed and lowered cost to the point where individual genes are sequenced routinely. The task now is to develop efficient methods for sequencing large regions of DNA .speed and lowering cost. Initially,

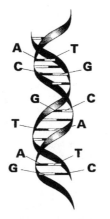

The four nucleotide bases are called A, T, C, and G for the first letter of each of their chemical names. In the double-stranded DNA helix, A on one strand always pairs with T on the other, and G always pairs with C. So, knowing the base sequence of one strand automatically reveals the sequence of the complementary strand. A and its complement T (A-T), or C-G, are therefore referred to as "base pairs."

this has been carried out primarily using the DNA of model organisms. The relative simplicity of genomes of certain animals and micro-organisms make them ideal terrain for technology development.

Genetic information in these organisms is simple and packed tightly together, so even when sequencing efficiency is not yet optimal for the human genome, it can still effectively mine valuable information from less-complex DNA.

So far, several laboratories have developed the capability to sequence at least 1 million base

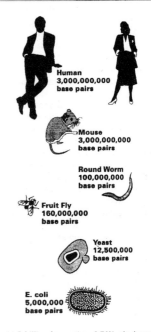

Human
3,000,000,000
base pairs

Mouse
3,000,000,000
base pairs

Round Worm
100,000,000
base pairs

Fruit Fly
160,000,000
base pairs

Yeast
12,500,000
base pairs

E. coli
5,000,000
base pairs

At 3 billion base pairs of DNA, the human genome is extremely large and complex. The Human Genome Project is also sequencing the genomes of important laboratory organisms to learn the most efficient methods for DNA sequencing.

pairs of DNA per year, at a cost of about $1 per base pair; a few can even do it for about 50 cents per base pair. Researchers sequencing the genome of a roundworm called *Caenorhabditis elegans* have increased their annual production rate to 15 million base pairs and have amassed more than 30 million bases of DNA sequence from that organism. These investigators expect to complete the sequence of the roundworm genome, which contains 100 million base pairs, by the end of 1998. The genomes of several micro-organisms, each containing from 1 million to 4 million base pairs, have been sequenced, and more will be soon. The complete sequence of the genome of baker's yeast was completed in the spring of 1996, an important step because yeast resemble human cells more closely than do bacteria.

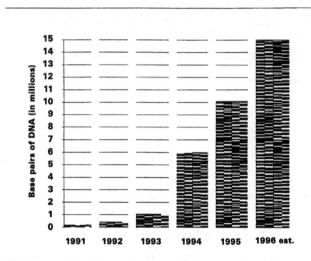

The amount of DNA that can be sequenced has increased dramatically since the Human Genome Project began. Researchers can now sequence 15 million base pairs in a single year.

Despite significant gains in sequencing capability, further improvements in DNA sequencing methods will be required to sequence the entire human genome. NCHGR has recently launched a new initiative to study test strategies for full-scale sequencing of mammalian DNA that would lead to the goal of sequencing the complete human genome by the year 2005. The outcome of these pilot studies will help scientists ramp up for sequencing human DNA in earnest.

Although providing a single reference sequence of the human genome will be an extraordinary accomplishment, further advances in sequencing technology are necessary so large amounts of DNA can be manipulated and compared with other genomes quickly and cheaply. Comparing differences in large regions of the genome--1 million bases or more--among

many individuals will yield an enormous amount of information about inheritance, such as disease susceptibility, response to certain environmental influences, and even evolution. So, while some researchers begin production sequencing of the human genome, others will continue to develop new methods for sequencing that will meet the ever-growing demands for cost-effective sequencing and comparison of very large genome regions.

The ultimate Human Genome Project task of sequencing all 3 billion base pairs in the human genome will provide scientists with a virtual blueprint of a human being. From there, researchers can begin to unravel biology's most complicated processes: how a baby develops from a single cell; how genes coordinate the functions of tissues and

organs; how disease predisposition occurs; and perhaps, even how the human brain works. Because this awesome task will require continued technology development and the participation of several new laboratories, NCHGR will increase its emphasis over the coming years on experimental sequencing of the human genome and innovation in technology.

DNA sequence derived by Human Genome Project laboratories is stored in databases scientists can freely access. In the United States, two databases, GenBank (http://www.ncbi.nlm.nih.gov), run by the National Center for Biotechnology Information, and the Genome Sequence Data Base (http://www.ncgr.org/gsdb), run by the National Center for Genome Resources, serve as public repositories of sequence information.

NCHGR Office of Communications, 31 Center Drive, Building 31, Room 4B09, MSC 2152, Tel: 301.402.0911, FAX: 301.402.2218
http://www.nchgr.nih.gov

F a c t s   a b o u t   D N A   S e q u e n c i n g

Part 1: Demographics

1. How old are you? _____ years of age

2. What is your current marital status?

   _____ Married

   _____ Widowed/divorced/separated

   _____ Never married

3a. Do you have any children?

   _____ No (skip to question 4)

   _____ Yes

   b. How old is your youngest child? _____ years of age

   c. Have any of your children been born with a birth defect or other congenital condition?

   _____ No

   _____ Yes

4. Is there any family history of genetic conditions or diseases in your family?

   _____ No

   _____ Yes

   _____ Not sure

5. Have you, or has anyone in your immediate family, ever received genetic counseling?

   _____ No, no one

   _____ Yes

   _____ Not sure

6. In which ethnic group do you classify yourself?

    _____ Hispanic

    _____ American Indian or Alaskan Native

    _____ Asian or Pacific Islander

    _____ African American

    _____ Causasian

    _____ Other (please specify:

7a. Have you ever attended any courses in human genetics?

    _____ No

    _____ Yes (please specify): _____

  b. Did you consider this course material relevant to nursing?

    _____ No

    _____ Yes

8. How thoroughly was genetics covered in other courses required for your nursing-education programs?

    _____ Not covered

    _____ Inadequately covered

    _____ Adequately covered

    _____ Very thoroughly covered

9. What are your main sources of information about new health problems and health care practices?

    _____ Professional journals

    _____ Professional associations

    _____ Nurses

    _____ Doctors

    _____ Other health professionals

    _____ Seminars and conferences

    _____ Continuing-education courses

    _____ Newspapers and news magazines

    _____ Other (please specify): _____

Part 2: Professional History and Goals

Place yourself on the scale with an **X,** *describing realistically your knowledge of human genetics (genetics nursing) or your application of human genetics.*

1. My experience in this area of practice has been

   _____

   None   Minimal   1–2 years   2–5 years   5–10 years   more than 10 years

2. My working knowledge of the current literature in this area is that I have

   _____

None	Minimal understanding	A broad overview	Read nursing journals	Read beyond nursing litera-ture into other disciplines	A comprehensive understanding

3. My comfort level in this area of nursing is

   _____

   Extremely low   Moderate   Totally at ease

4. The importance of incorporating genetic skills and knowledge into the work that I do is _____
   Minimal   Some   A lot   Tremendous

5. By reading this book, I hoped to learn as much as I could about

   a. _____   c. _____

   b. _____   d. _____

6. This book will help me in my work by _____

   _____

7. My personal goal for reading this book was _____

   _____

   I did/did not achieve this goal. *(Please circle response.)*

8. Having finished reading this text, I will feel I have accomplished an important achievement if I can _____

   _____

   _____

Part 3: Nursing Education in Genetics: Topics Review

The purpose of this section of the self-assessment is to help you to clarify your views on topics that are important to nursing education in genetics. This survey is meant to be completed just after reading the text to examine your views and assess whether and how they have changed since you completed the assessment at the front of the book.

1. Below is a list of topics that are included in the text; for each topic, please indicate (by circling the number that best describes) whether you consider it

    1   Very important
    2   Somewhat important
    3   Not very important
    4   Not at all important
    5   Not sure or don't know

    a. Advantages and disadvantages of participating in family
        studies of genetic conditions   1   2   3   4   5

    b. Who has access to genetic information   1   2   3   4   5

    c. Screening of the general public for genetic disorders   1   2   3   4   5

    d. Treatments for genetic disorders, including
        gene therapy   1   2   3   4   5
    e. Including children in decisions about their
        genetic testing   1   2   3   4   5

    f. Coping with a new genetic diagnosis in the family   1   2   3   4   5

    g. Genetic disorders and health insurance   1   2   3   4   5

    h. Genetic information and employment   1   2   3   4   5

    i. Genetic testing and biotechnology companies   1   2   3   4   5

    j. New genetic information and the legal system   1   2   3   4   5

    k. How to stay informed about new developments in the
        Human Genome Project   1   2   3   4   5

l. The media and interpreting the results of genetic
    research   1   2   3   4   5
m. How society may be affected by the Human Genome
    Project   1   2   3   4   5

n. Health care reform and the Human Genome Project   1   2   3   4   5

o. Family and professional partnerships   1   2   3   4   5

p. Basic human genetic concepts   1   2   3   4   5

q. Single-gene inheritance   1   2   3   4   5

r. Components of a family history   1   2   3   4   5

s. Molecular genetic clinical applications   1   2   3   4   5

2. Are there issues or topics related to genetic education for nurses that were not covered in this book but that you consider important?

_____ No

_____ Yes (please comment below)

Comments: _____

_____

_____

_____

_____

_____

_____

_____

GLOSSARY

acquired mutations gene changes that arise within individual cells and accumulate throughout a person's lifetime; also called *somatic mutations.*

allele an alternative form of a gene or DNA segment. An allele for each gene is inherited from each parent.

amino acid any 1 of 20 nitrogenous molecules that are linked in a linear sequence to form proteins. The sequence of amino acids determines the protein formed and its function.

apoptosis the normal progress of cellular death. This pathway may be disrupted, leading to malignant transformation of cells.

assessment the systematic process of collecting relevant patient data for the purpose of determining actual or potential health problems and functional status. Methods used to obtain data include interviews, observations, physical examinations, review of records, collaboration with colleagues, and consideration of applicable literature and research.

autonomy the right to make personal decisions about one's health and property, which is consistent with the importance of individual liberty in Western culture and the Western legal system (self-determination).

autosome a single chromosome from any 1 of the 22 pairs of chromosomes not involved in sex determination (XX or XY).

base pair two paired nitrogenous bases, either adenine and thymine or guanine and cytosine, held together by hydrogen bonds.

base sequence the order of nucleotide bases in a DNA molecule.

beneficence an ethical principle, central to biomedical ethics, that promotes doing good in all actions.

birth defect abnormal congenital condition, ranging from minor to severe, that may result in a debilitating disease, a physical or mental disability, or early death. Birth defects may or may not have a genetic cause.

carcinogenesis a multistep process resulting from the accumulation of numerous genetic mutations leading to uncontrolled cellular growth.

carrier term used to refer to any individual who carries a single copy of an altered gene (mutation) for a recessive condition on one of a chromosome pair, and a copy of the "normal" form of that gene on the other chromosome. People who are carriers of a gene mutation have a 1 in 2 (50%) chance of transmitting the mutation to each offspring but do not exhibit the condition themselves. Each person in the general population is a carrier of, on average, 5 or 6 gene mutations for recessive disorders.

cell a small, membrane-bound compartment containing chemicals (including DNA) needed for proper body functioning.

chromosomes structures found in the nucleus of a cell that contain the genes. Chromosomes occur in pairs, and a normal human cell contains 46 chromosomes, 22 pairs of autosomes, and 2 sex chromosomes.

confidentiality communicated, conveyed, acted on, or practiced in confidence, private, shared only with a limited few.

congenital present at birth.

consanguinity related by having a common ancestor; close blood relationship.

crossing over process that occurs during meiosis, in which homologous maternal and paternal chromosomes break and exchange corresponding sections of DNA, and then rejoin. This process can cause an exchange of alleles between chromosomes.

cytoplasm the cellular substance outside of the nucleus of the cell in which other cell components are suspended.

DNA the chemical compound deoxyribonucleic acid, located primarily in the nucleus of all cells. DNA carries the instructions (genetic code) for making all of the proteins and other materials needed for growth, differentiation, and development of an organism.

DNA replication the use of existing DNA as the pattern, or template, for the creation of new DNA strands. In humans, replication occurs in the cell nucleus.

DNA sequence the relative order of base pairs that occurs in DNA, a gene, or a chromosome.

dominant a genetic trait that is expressed when a person has a gene mutation on one of a pair of chromosomes and the "normal" form of the gene on the other chromosome. A person who has a dominant gene usually expresses the trait.

double helix the shape of two strands of DNA, similar to a spiral staircase.

ELSI Ethical, Legal, and Social Implications Program of the Human Genome Project.

eugenics the use of genetic knowledge and technology to alter population characteristics (i.e., to create "genetic perfection" by promoting selective reproduction).

exons the part of the DNA sequence that codes for a protein (compare to *introns*).

gamete a male or female reproductive cell (egg or sperm), which contains 23 chromosomes rather than the usual 46 found in the rest of the body.

gene a functional unit of inheritance consisting of DNA.

gene mapping determination of the relative positions, distance, and linkage of genes on the chromosome.

gene sequencing determining the exact order of the base pairs in a segment of DNA.

gene testing analysis of a sample of blood or other body tissue for biochemical, chromosomal, or genetic markers that indicate the presence or absence of a genetic condition or predisposition.

gene therapy a method of treating genetic disorders by manipulating the genes themselves.

genetic code the sequence of three bases, called *base triplets,* that specify the 20 amino acids found in proteins.

genetic condition variations, disorders, birth defects, or diseases that are caused or influenced by genes and that may or may not be transmitted from parent to offspring.

genetic diagnosis cytogenetic, biochemical, or molecular testing or identification of a clinical phenotype that identifies the patient as having a genetic condition.

genetics the scientific study of heredity; how specific traits or predispositions are transmitted from parents to offspring.

genome all of the genetic material contained in the chromosomes of a particular organism.

genotype the genes and the variations therein that a person inherits from his or her parents.

germ cell a sperm or egg, or any cell that becomes a sperm or egg.

germline mutation any gene alteration that occurs in the body's reproductive (germ) cells and that can be passed on to future generations; also referred to as an *hereditary gene mutation.*

heterozygote a person who has different alleles at a given location on a specific chromosome.

Human Genome Project an international research effort aimed at identifying and characterizing the order of every base in the human genome.

informed consent a communication process between a health care provider and a patient, characterized by mutual participation, respect, and shared decision making, the goal of which is to actualize patient choices, including the right to accept or refuse treatments.

introns the DNA base sequences that interrupt the protein coding sequences (exons) of a gene. Introns are transcribed into RNA but are removed from the message before translation into a protein.

linkage term used to describe genes that are located relatively close to one another on the same chromosome and that are inherited together.

locus the position on a chromosome of a gene or DNA marker.

meiosis the process of cell division that produces reproductive cells (egg or sperm). Meiosis results in daughter cells, which contain half of the chromosome complement (23).

messenger RNA   RNA that serves as a blueprint for protein synthesis (translation).

mismatch repair genes   genes controlling the accuracy of DNA replication during cell division.

mitosis   the process of cell division that results in daughter cells that are genetically identical to each other and to the parent cell.

multifactorial   health conditions determined by multiple factors, including genetic and environmental factors, each having its additive effects.

mutation   any change in DNA that alters a gene, which can produce deformity or disease or have a neutral or beneficial effect. Mutations occur spontaneously during cell division or may be caused by environmental influences, such as radiation or chemicals.

nitrogenous base   a nitrogen-containing molecule that has the chemical properties of a base. In DNA, these are adenine, thymine, guanine, and cytosine. In RNA, uracil is substituted for thymine.

nucleotide   a small unit of DNA or RNA consisting of a nitrogenous base (adenine, thymine, guanine, cytosine, and, in RNA, uracil), a phosphate molecule, and a sugar molecule. Thousands of nucleotides link together to form a DNA or RNA molecule.

oncogenes   genes that normally play a role in the growth of cells but, when overexpressed or mutated, can foster the growth of cancer.

pedigree   also referred to as a *family tree*; a pictorial family history diagram that traces genetic characteristics and disorders in a family.

phenotype   a person's entire physical, biochemical, and physiological makeup, as determined by the individual's genotype and environmental factors.

physical map   a map of the locations of identifiable landmarks on DNA, including genes and gene marker sites.

polymorphism   differing DNA sequences (genetic variations) among individuals. Polymorphisms in populations are often useful for creating genetic-linkage maps.

predisposition   having an increased susceptibility to a health condition such as cancer, as determined by genetic analysis.

premutation   a change in DNA that has the potential to lead to gene alterations in future generations. People who have premutations do not usually exhibit the condition.

presymptomatic   not yet showing the signs and symptoms of a disease.

recessive   a genetic trait that is expressed only when a person has two copies of a mutant autosomal gene or a single copy of a mutant X-linked gene in the absence of another X chromosome.

recombination   also referred to as *crossing over*; the process by which genes are exchanged between parental chromosomes and then passed on to offspring.

reproductive cells egg and sperm cells (gametes), containing a single set of chromosomes (23).

RNA ribonucleic acid, a chemical similar to DNA but differing in the type of sugar (ribose instead of deoxyribose) and the substitution of uracil for thymine as the complement to adenine. Several types of RNA play important roles in protein synthesis.

sequencing determining the order of nucleotides or base pairs in a DNA or RNA molecule or the order of amino acids in a protein.

sex chromosomes the set of chromosomes that determine the gender of an organism. Human females have two X chromosomes; human males have one X and one Y chromosome.

somatic cells any cell in the body, excluding reproductive cells.

telomere the extreme end of each chromosome arm.

trait an observable physical property or characteristic of an organism, usually used to refer to a nonmorbid characteristic, such as eye color.

transcription the process of transforming information from DNA into new strands of messenger RNA.

translation the process of carrying out instructions from the messenger-RNA triplet code into chains of amino acids, which then become proteins. This process takes place in the cytoplasm.

translocation the joining of a part of, or a whole, chromosome to another.

trisomy having three chromosomes instead of the usual two (e.g., trisomy 21).

tumor-suppressor genes genes that normally restrain cell growth but, when missing or inactivated by mutation, allow cells to grow uncontrolled.

X-linked genes located on the X chromosome. One altered gene on an X chromosome in a male can produce disease, such as hemophilia.

zygote a fertilized egg.

INDEX

ACOG. *See* American College of Obstetrics and Gynecology
Adenomatous polyposis coli (APC) gene, 158
AFP screening. *See* Alpha-fetoprotein screening
Alleles, 4, 31
Alliance of Genetic Support Groups, 89
Alpha-fetoprotein screening, 87, 104
Alzheimer's disease, 246–248
American Board of Genetic Counseling, 18–19
American College of Medical Genetics, 16, 153
American College of Obstetrics and Gynecology, 79
American Society of Human Genetics, 16, 18, 110, 159–160
Amino acids, 29–30
Aneuploidy, 53–54, 185
Angelman syndrome, 60
Apoptosis, 184
ASHG. *See* American Society of Human Genetics
Autonomous decision making, 228, 241, 246–248
Autosomal
  -dominant inheritance, 43, 45, 46, 47, 48, 49, 61–62
  -recessive disorders, 46–47, 48, 49, 62, 79
  -recessive inheritance, 43, 46–47, 79
Autosomes, 31

Base pairing, 29
Beneficence, 224
Bloom's syndrome, 187–188
Breast cancer
  BRCA1 gene, 215, 216, 217, 220
  predisposition to, 215, 216

CAG repeat, Huntington's disease, 154
Cancer genetics
  causes, 181–182
  cell properties, 179

chromosomal abnormalities, 184–185
  familial, 121, 123, 125, 128, 181–182, 197, 198–203
  research, 188–189
  risk assessment, 195–196
  risk-counseling process, 198–203
    hereditary nonpolyposis colon cancer, 182, 184, 199
  susceptibility testing, 203–210
Cancer-causing genes and anomalies
  BRCA1 and BRCA2, 215, 216, 217, 220
  chromosome abnormalities, 184–185, 187–188
  classes of, 182–184
  gene amplification, 185–186
  gene deletions, 187
  genomic imprinting, 188
  mutations, 186–187
Carcinogenesis, 179
Care ethic, 224, 241
Carrier screening, risks and benefits, 87–88
Carrier testing
  for autosomal-recessive disorders, 79
  for cystic fibrosis, 105, 147
  for fragile X syndrome, 153
  informed consent for, 87–88, 105
Case studies
  autonomy in decision making, 246–248
  carrier testing for cystic fibrosis, 105
  disclosing breast-cancer research results, ·216–217
  Down syndrome, 68–70
  DNA research testing with linkage analysis, 107–108
  Duchenne muscular dystrophy, 65–66
  familial adenomatous polyposis, 171–175
  family notification responsibility, 218–219
  follow-up after genetic diagnosis, 142–143
  fragile X syndrome, 165–168
  genetic health history assessment, 102–103